William Porter

Life, and its Forces

Health and Disease Correctly Defined

William Porter

Life, and its Forces
Health and Disease Correctly Defined

ISBN/EAN: 9783744738255

Printed in Europe, USA, Canada, Australia, Japan

Cover: Foto ©berggeist007 / pixelio.de

More available books at **www.hansebooks.com**

LIFE, AND ITS FORCES.

HEALTH AND DISEASE

CORRECTLY DEFINED.

A RELIABLE GUIDE TO HEALTH WITHOUT THE USE OF
MINERAL OR VEGETABLE POISONS, OR IRRITANTS.

THE CONCLUSIONS FROM FORTY YEARS' PRACTICE
OF MEDICINE.

BY

DR. WILLIAM PORTER,

311 FAIRFIELD AVE., BRIDGEPORT, CONN.

HARTFORD, CONN.:
PRESS OF THE CASE, LOCKWOOD & BRAINARD COMPANY.
1878.

INDEX.

MEDICINES.

1*

EXORDIUM.

Awake from the slumber of ages!

Purify your physical bodies, that they may be a fit dwelling-place of a good spirit.

For the good time which has been so long expected, will surely come, when by progress of mind in wisdom, it is compelled to come.

" Man shall not live by bread alone, but by every word which " he is sufficiently developed to hear.

Superficial calleth unto superficial, and " deep calleth unto deep;" and the answers give wisdom according to the capacity of the interrogator to receive.

And the individualized spirit has power to direct the forces of existence for good or for evil purposes, according to its attainment of wisdom in harmony with unchangeable laws.

Each individual has within itself the power of blessing or cursing, according to the capacity of its organism, and its greatest love in determinate action ; and also according to harmony with divine intelligence or its opposite.

INTRODUCTION.

THE highest degree of healthy sensitiveness of the nervous system is a necessary condition to receive the highest inspiration from divine wisdom, which is so indispensably necessary for the elevation of the human race from ignorance and its consequences.

During the past three or four thousand years many theories have been invented, each claiming to explain concerning disease and remedies; but each was soon superseded by others equally at fault, though often antagonistic to previous ones. The practice founded on these false theories has been more destructive to the human race than any other cause. not excepting war and pestilence.

Considering the general ignorance of men and women concerning physical health and disease, all attempts to agitate thought on the subject seem to be overwhelmingly hopeless; and were it not for the fact that " universal mind " has been forced to dread what is popularly termed medicine ; and also if I had not faith of larger size than " a grain of mustard seed " in the omnipotence of truth and certainty of progress, I should not have attempted to write these few pages.

Such general ignorance on so important a subject is a natural consequence of depending on a class called Doctors, whose financial prosperity depends on the people remaining ignorant and diseased.

I would that I could speak in thunder tones to engage the requisite attention to effectually warn every individual to stop at once and for ever the highly pernicious practice of using narcotics and other poisons to destroy sensibility by which to relieve pain.

But, says one, " I had a pain, and by the advice of a celebrated doctor I took a dose of morphine and it relieved me." I answer that there was a cause for that pain which the morphine did not remove; but it destroyed the force which made it possible for a telegraphic despatch by the nerves giving information of obstruction to the circulation, so that the cause could be removed.

A few more victories of that kind would amount to a disastrous defeat. A restless child can sometimes be quieted by the same means or principle, but its senses will be so deadened that it will grow up foolish just in the ratio that its susceptibility has been destroyed. The world is already burdened with those whose intellects have been dwarfed in childhood by narcotics.

I do not propose in this work to give a list of all articles which may be good for medicine, as I do not intend to write so much that it will never be sufficiently understood to be useful as a guide.

I have selected a few of the best means yet discovered to remove obstruction, thus working directly on the more evident causes of disorder : and as they harmonize with a healthy state, their good effects can be confidently relied upon. if used with common prudence and judgment.

I desire particularly to urge the necessity of understanding correct principles involved in health and disease, as they apply to all conditions : thus avoiding the supposed necessity of selecting a name from among from two to three thousand species of disease, with their peculiar symptoms ; also knowing that two cases were never precisely alike, as that is a plan by which the allopaths have stultified themselves ; and its utter impracticability has disgusted and discouraged many reasonable minds from following it as a profession.

A vast amount of brain-work has been applied to observe and classify the symptoms of disease ; the record of it has been the burden of many published volumes ; and a vast number of theories have been invented how to destroy symptoms, but nearly all the progress that has been made during the past ages is summed up in the fact that a greater potency overcomes a less. And as most of the means used have been pernicious and deadly in their tendencies, it corresponds to the wisdom of burning a barn to kill the rat in it. The term medicine should never be applied to any substance which is not in perfect harmony with a healthy condition.

1 am indebted to Dr. Samuel Thomson, the founder of the system of practice which bears his name, for many ideas contained in this work, having practiced

under his supervision in 1841 and 1842 : indeed, the medicines recommended are from among the number of those which he selected, and I can, after using them forty years, add my testimony to his, that they can be relied upon as efficient, and in harmony with health.

I am confident that in almost every family, certainly in every neighborhood, there are individuals who have capacity and good judgment sufficient to understand and practice the instructions which I have given to successfully treat any curable case of disorder, where surgical skill is not needed. If this work is instrumental in agitating thought on the subject of health and disease sufficient to diminish the amount of human suffering and misery even in a small degree, it will be a source of gratification to the author.

W. P.

BRIEF OUTLINE OF THE PRINCIPAL ORGANS OF THE HUMAN BODY.

A very brief notice of a number of the principal organs of the human body, and their functions, may not be out of place in this work, as a profound knowledge of anatomy or physiology is not necessary for a good understanding of the nature of remedies for disorder.

In the rudimental condition or first stages of formation of the human body, the embryotic heart is manifest by its motion, and it has been called "the first part that moves and the last that dies," of the human machine.

THE SKELETON, or framework, is composed of two hundred and forty bones, of which there are sixty-three in the head, including thirty-two teeth; fifty-three in the trunk; sixty-four in the upper, and sixty in the lower extremities.

The bones are covered with a membrane, which contains the blood-vessels by means of which they are nourished. In health the bones are insensible, but when diseased they are capable of great pain.

THE MUSCLES of the human body are mostly in pairs; they constitute the flesh, and consist of fibers which in bundles are enveloped in sheaths, and sup-

2

plied with arteries, veins, lymphatics, and nerves; the muscles being fastened to the bones, produce, by means of Life's forces acting on the nerves, the expansion and contraction which move the whole body.

TENDONS unite the bones with the muscles.

LIGAMENTS are· the cords or bands which connect and keep the joints and other organs in their places.

CARTILAGE, or gristle, unite bones together and cover the ends of those which form joints. The front part of ribs consists of cartilage.

TISSUES OR MEMBRANES.—The mucous membrane is the lining of the nose, mouth, throat, lungs, stomach, intestines, vagina, urethra, and other organs. It secretes a peculiar slimy mucus sufficient to lubricate the membrane when in health. In disease, especially when perspiration is checked, it secretes an inordinate quantity, which, when thick, is denominated phlegm; when viscous, it has been called false membrane. It forms in the windpipe in croup; it, however, has no vessels of circulation, and is simply an exudation of waste or morbid matter, the result of the excretive process being checked. The mucous membrane is subject to inflammation, tuberculation, ulceration, and when inflamed or covered with viscous secretion, the blood-making process is hindered more or less, and more or less weakness is the result. It is also subject to bleeding when much irritated or inflamed; it can be much disordered with but little, if any, uncomfortable feeling from it, not being directly supplied with nerves of sensation; but if the disorder or inflammation extends to the outward covering. or serous membrane, where nerves of sensation are distributed, pain may be felt. The mucous surface in health secretes more or less an alkaline substance.

THE SEROUS MEMBRANE.—This covers the brain, lungs, stomach, intestines, and other organs, and also lines the chest and abdomen. It exhales a serum, or watery matter, which in undue quantity, if not absorbed and excreted, remains as dropsical accumulations. The organization of the human body corresponds in many respects to a galvanic battery, the serous and mucous surfaces corresponding to the plates, and the nerves as conductors.

THE CELLULAR MEMBRANE is formed of many cells, which communicate with each other and fill the spaces between the skin, muscles, and other solid parts; the cells are moistened with watery matter, which, if in excess and not excreted, remains as dropsy.

THE SKIN is the external covering of the body; it is plentifully supplied with nerves and vessels of circulation. also with pores, or excretory channels for elimination of waste matter. It is of the utmost importance that the healthy functions of the skin should be uninterrupted, for checked perspiration is the evident commencement of disorder generally. When perspiration through the skin is checked, the mucous membrane begins to perspire or sweat, and mucus accumulates in the stomach and intestines and seriously hinders digestion of food.

It has been estimated from the number of pores

opening on the external surface of a square inch of skin, that a person of ordinary size would be furnished with about seven millions; with an aggregate length of from twenty-five to twenty-eight miles, through which the blood should be relieved of much of its waste or worn out matter by perspiration.

THE GLANDS are parts composed of arteries, veins, and absorbents, and perform the work of separating certain fluids from the circulation; they are very numerous and perform different offices: as the liver, to secrete bile; the female breast, to secrete milk: the salivary glands, etc., etc. The lymphatic glands often enlarge and form tumors.

THE LIVER is the largest gland of the body; it extends from the lower part of the diaphragm of the right side, to a little past the center of the body into the left side. One of its offices in health is to secrete bile from the blood.

It collects in the gall-bladder and discharges its contents into the duodenum or second stomach. The average weight of the liver is between three and five pounds, but may be, by disease, increased to twenty or thirty pounds. Its connection by the nerves with the spine is in the intervertebral space at the right side, between the seventh and eighth dorsal vertebra, and nearly on a level with the lower part of the right shoulder-blade.

THE SPLEEN is a soft, spongy organ situated above the left kidney in the upper part of the abdominal cavity. It is connected with the stomach by the cellular membrane and by small blood-vessels, and is attached to the lower part of the diaphragm. It has no special excretory duct : its nervous communication with the left side of the spine is in the intervertebral spaces between the seventh and ninth dorsal vertebra.

THE PANCREAS is a glandular body situated behind the stomach and across the spine, in contact with the spleen, and having an excretory duct which opens into the duodenum.

THE STOMACH is a membranous sac, into which the food passes for the digestive process. It is located in the upper part of the abdominal cavity, and its capacity varies, when distended, from a half pint to a gallon or more. It is composed of three layers or coats : the external or serous surface ; the middle or muscular : and the internal or mucous. The stomach is situated below and in contact with the diaphragm on the left side. The food passes into the stomach through the gullet or esophagus. a little to the left of the center. The outlet is the pylorus. at the right side or small extremity of the stomach, where it empties its contents into the duodenum or second stomach. The left extremity is connected with the spleen. The stomach is connected with the spine by the nerves in the intervertebral spaces between the second and fourth dorsal. The duodenum or second stomach is by the nerves connected with the spine in the spaces between the fifth and sixth dorsal vertebra.

THE INTESTINES are that portion of the alimentary canal which extends from the outlet of the stom-

2*

ach to the anus, and are estimated to be five or six times longer than the body. Millions of absorbent vessels open on the internal surface of the stomach and intestines, whose office is to absorb the products of the digestion of food for the blood-making process.

The small intestines are connected by the nerves with the spine in the intervertebral space between the eleventh and twelfth dorsal vertebra.

THE URINARY ORGANS are the kidneys, ureters, bladder, and urethra. Two glands, the kidneys, placed each side of the spine, and in contact with the lower part of the diaphragm, secrete urine from the circulation which is poured from their inner cavities through the ureters into the bladder, and from thence excreted through the urethra. The right kidney is connected by the nerves to the space between the twelfth dorsal and first lumbar vertebra at the right side of the spine. The left kidney is connected with the space between the left side of the twelfth dorsal and first lumbar.

THE URETERS are membranous tubes leading from the kidneys to the bladder.

THE BLADDER is a receptacle of the urine, and is placed between the pubes or front bone and the rectum in the male, and between the pubes and the vagina in the female. The nervous connection of the bladder with the spine is in the intervertebral space between the twelfth dorsal and first lumbar.

THE URETHRA is a membranous canal leading from the neck of the bladder, by which its contents are passed from the body.

The genital organs are also connected by the nerves to the space between the fifth lumbar vertebra and the os coccyx.

THE ABSORBENTS.—The lymphatic vessels are very numerous, permeating or having their rise in every part of the body : with the lacteals they constitute the absorbent system. They absorb fluids from different parts of the body : from the mucous and serous surfaces, including the skin. After passing through an extensive system of glands, they pass their contents into the blood. except what is not excreted by perspiration.

THE LUNGS are spongy bodies filling the cavity of the chest, and connect with the passages from the mouth and nose by the windpipe. They are divided into two portions by a partition called the mediastinum. which is attached to the spine and breast bone. The right lung is the larger, and consists of three lobes. The left lung consists of two lobes, between which the heart is situated. The bronchi are the names given to the branches or divisions of the windpipe which lead the air into and out of the air cells. The lungs are connected by the nerves with the spine in the spaces between the seventh cervical and first dorsal vertebra.

THE DIAPHRAGM is a large, flat muscle, dividing the chest from the abdomen. When the lungs are inflated it is nearly level, but when the breath is thrown out it is arched·upward. Its motion is governed to a great extent by the will through the motor nerves. It is an important agent in respiration.

THE HEART, AND CIRCULATION OF THE BLOOD.—
The heart is a strong muscle, situated in the left cavity of the chest; it is a double organ resting upon the diaphragm, and is nearly covered by the lobes of the left lung. It is divided into four cavities: the right auricle receives the blood from the veins and absorbents through the vena cava ascendens, from below; and through the vena cava descendens, from above: the blood then passes to the right ventricle, from which it then passes through the pulmonary artery to its branches; one to the right, the other to the left lung. It then permeates the capillary vessels which surround the air cells where it receives the electricity from the oxygen of the air, which gives it a bright, scarlet color, and also parts with some of its waste matter, consisting, in part, of carbonic acid gas and water. It then returns by the pulmonary veins to the left auricle; from thence to the left ventricle, and from there it passes to the aorta, which supplies by its branches red blood to every part of the body.

In passing through the capillaries it parts with its nutritive properties and bright scarlet color, and is collected by and returns through the veins to the heart and lungs, to be again vitalized by electricity of oxygen, and distributed as before.

THE BRAIN AND NERVOUS SYSTEM.

The brain is a mass of nervous matter contained in the skull, and is the grand central telegraphic depot where messages are received by means of the nerves, and from whence messages are sent either by direction of the individual's will, or by the sleepless indwelling life which directs and continues the process of digestion, absorption, and peristaltic motion, inspiration and respiration, when the individual is asleep or unconscious.

The spinal marrow is a continuation of the brain, and appears to be a grand trunk conductor, which is divisible into the cerebro-spinal, or part pertaining to both the cerebrum and spine: the cervical, or part pertaining to the neck: the dorsal, pertaining to the back: the lumbar, pertaining to the loins: the sacral, pertaining to the sacrum, or bone between the hip bones, supporting the spinal column: and the coccygeal, or the extremity of the spinal column.

The cerebro-spinal part furnishes nerves to the eyes, ears, nose, face, tongue, throat, teeth, and muscles of the neck. Also, the pneumogastric nerve from the medulla oblongata, or top of the spine, furnishing communication with the mouth, throat, lungs, stomach, heart, liver, spleen, kidneys, and bowels, membranes and muscles.

The great sympathetic nerve consists of a double chain of ganglia running down each side of the spinal column: it is composed of motor and sensitive filaments, and is distributed to organs over which the will has no direct control, as the heart, liver, kidneys, etc. It also accompanies every artery and blood-vessel to their minutest extension.

A current through the nerves of sensation proceeds from without, inward through the gray substance of the brain : sensibility is destroyed by division of the gray substance.

The impulse which gives rise to voluntary motion is transmitted by the brain through the white substance of the anterior and lateral columns of the spine from within outward. The power of causing motion is destroyed by division of the white substance.

The motor nerves at the medulla oblongata cross to opposite sides ; therefore, injury of one side of the brain may produce a paralysis of the opposite side of the body.

Throughout all nature or the universe life is a unit consisting of innumerable individualities or forms, through which its manifestations are perceived by individuals by the nerves of sensation. It is difficult for the human mind to conceive the infinite capacity and susceptibility of the nervous system in a healthy condition. We will look at a good photograph of any object and perceive the delicacy of the lights and shadows ; we will remember that the vibrations of rays of light must be almost infinitely varied to produce that delicate shading : also, as we look at it, the same infinite variety of vibrations are necessary to impress the picture on the retina of the eye ; now, if several persons look at it at the same time, the same unchangeable law is in full force between it and all the observers.

We may say that peculiar modifications of life's forces has been, by the direction of mind, permanently impressed on that paper, with its peculiar emanations, to remain until obliterated by a more powerful and positive impression.

A vivid description of a disgusting fact is sufficient to cause some negative and sensitive person to vomit. The positive and peculiar vibrations called sound were sufficiently effectual to impress disagreeably the nerves of sensation, and the individualized spirit directed force to repel and throw off the impression.

Thus everything conceivable is a cause, and has its inevitable and peculiar dynamic qualities, which it is in the power of mind, according to its attainment of wisdom to direct and govern.

LIFE, AND ITS FORCES.

SECTION 1. Life : the all-mighty in all things.

SEC. 2. Life and intelligence never separate; individualized in forms ; each a part mighty.

SEC. 3. We know nothing of life except by its manifestations, which are limited by the capacity of the organization by which it manifests.

As individualizations of life and intelligence in human forms, we may scan life's forces and perceive the method of working of life in motion.

SEC. 4. Life's forces may be named electricity, and its innumerable modifications.

SEC. 5. In equilibrium it is all-pervading.

SEC. 6. As energized and directed with sufficient velocity through a resisting medium, we have heat, or fever, as one of the results.

SEC. 7. Circulating through an unequally resisting medium, polarity is induced, which is termed magnetism ; with qualities positive and negative ; the positive possessing an expansive, the negative a contractive force.

Sec. 8. Magnetic poles of the same denomination repel, but opposite denominations attract each other.

Sec. 9. Attraction, or love, and repulsion, or hate, the two extremes; including the intermediate law of equilibrium or passivity.

Sec. 10. The innumerable modifications of life's forces act on each other for good, bad, or indifferent, according to their adaptability.

Thus, phosphate of lime pulverized from the rock, and phosphate of lime pulverized from animal bones, are chemically the same; but the dynamic emanations from the first as plant food are worthless, from their non-adaptability; while the latter is an excellent fertilizer, being more refined and better adapted to the growth of vegetable forms.

Sec. 11. By patiently and critically observing the laws by which life's forces are governed, their attractive and contractive, their expansive and repulsive powers, we are enabled to gain an insight into the mode of operation of formation and decomposition of forms, and learn how to promote an orderly and change a disorderly action of them.

Sec. 12. An atom, or particle, of what we name matter, radiates an electric influence of a certain modification exactly in accordance with the quality of the atom; and when an atom or particle is within the sphere of influence of another atom whose emanations are different in velocity of circulation, a current becomes determinate from one to the other, or from the positive to the negative, according to the laws of

3

equilibrium, and also of polarity. And if the attraction is sufficient they come together and are in combination by those laws, or the law of attraction, or love.

SEC. 13. When united long enough for the particles to become in a state of perfect equilibrium, or of one polarity, and it may be a short time or a work of ages, the particles repel each other, and disintegration ·
commences. Thus formation and decomposition, or life in motion, is continual.

SEC. 14. Particles by those laws become aggregated in various forms, and as life and intelligence exists, and is constantly active, change is the result, else all would be death, which is not an existence. And the various forms of what we name matter, becomes a means by which life is manifested invariably in accordance with the capacity of the organism : and as life and intelligence are never separate, individuality of intelligence, or spirit, has its birth.

Thus an aggregation of matter in the form of a tree, manifests enough intelligence to form new channels of circulation to repair damages when the bark is cut, and the laws of individual rights are observed in a forest in regard to space in which to grow.

A well-developed, symmetrical human body, in its best or unobstructed condition, is of capacity for the highest manifestation of life of any form on this planet, and is a fit dwelling-place for a good or holy spirit ; but certainly not if the dwelling becomes disgustingly filthy by waste or worn-out matter being ·

retained, or by being saturated with tobacco, whiskey, or poisons of any kind which have been very improperly termed medicines.

After individualization of spirit by organization of matter in form, the spirit, according to its capacity as an individual, perceives the necessity of controlling, regulating, and keeping in proper condition the physical organism, so that it will be a good servant for use. And it will invariably do it according to its wisdom : and if it was independent it would soon be without a physical body as a means of development, but being more or less magnetically connected with everything else, and subject to the law of equilibrium and polarity, what is wanting is attracted, and what is in excess is repelled to a great extent ; so that a renovating process is active while the individual is asleep, or is not positive, but in a sufficiently passive and receptive condition ; thus being connected with the Great Spirit or Life of the Universe, which is— Unity.

SEC. 15. There are two prominent principles by which life manifests itself in constant activity, viz. : the positive and negative, or male and female forces, prominently manifested in visible changes of matter, and through man as positive, and woman as negative, (the reverse condition of men and women appears to be disorderly exceptions,) and no fact is better established than that the evolutions of life as observed in offspring of individuals, that such offspring are almost invariably the representative of the spiritual states or qualities of the individuals which produced them.

SEC. 16. If men and women would generate and cultivate good, loving, harmonious, and healthy children, with mental capacity well able to be useful as well as ornamental to the world, they can do so, provided they are of that quality themselves. But if they are quarrelsome, inharmonious, fickle-minded, intemperate, knavish, inordinately sensual, and diseased themselves, they need not be surprised if their children are of similar quality, for they are, in fact, the offspring of the spiritual conditions of both the parents, but materially modified by the condition of the mother during gestation, and later by impressions during childhood and youth by education. There also appears to be a few exceptions to general laws. The physical body is a good servant for manifestation of the individualized spirit just in accordance with the perfection, quality, and condition of the organism.

A HEALTHY CONDITION.

SEC. 17. A healthy, or unobstructed state, is by good digestion of wholesome material in the stomach, and sufficient inspiration of good air in the lungs, so that sufficient force and heat is evolved to fully expand and keep the whole system in a free, unobstructed, breathing condition; not only an inspiration and expiration by the lungs, but by every pore of the body opening on the external surface; a state which generates or liberates a sufficiency of forces by which the blood is circulated through the arteries, and by means of the capillary vessels nourishes the whole body, and then returns it by the veins.

A human body in a healthy, unobstructed condition, is full of life's forces in activity and strength. Laziness, or weak nerves, are not thought of as an experience. Despondency, pains, aches, and bad feelings, are simply impossible. A very pleasant sensation of magnetic warmth and strength is felt in every part of the body, and existence is felt as a great pleasure; but the most prominent feeling is a sense of excess of energy, and it is a constant source of satisfaction to make use of it. It is Life, in the fullness of its manifestation according to the capacity of the organization of the form.

CAUSES AND PROGRESS OF DISEASE.

SEC. 18. It has been demonstrated that of the food and drink taken into a healthy stomach, a certain proportion of the products of digestion is taken by the absorbent vessels which open on the mucous surface of the stomach and intestines to form blood, while nearly two-thirds pass off through the external surface by sensible or insensible perspiration, and nearly one-third by the kidneys, the bowels, and the mucous surfaces, including the lungs. The imponderable or electric forces which are adapted are principally utilized to circulate the blood, to balance the pressure of the atmosphere, to be directed by the indwelling spirit through the involuntary nerves to perform the

3*

necessary functions of the system when awake or asleep, and also to be directed by the will through the voluntary nerves to expand or contract the muscles as a means of power to produce motion—effective according to the amount of such stock on hand.

Thus it is of the utmost importance that the excretionary process should not be checked, as it is a direct means by which the waste matter is eliminated; which, if retained in the system, is not only obstruction, but is evolving unadapted modifications of force which directly produces disorder.

Life's forces are governed to a great extent by the action of the mind, and when by sluggish inactivity not enough force is generated, or by too great activity more is dissipated than is supplied by the natural sources of supply for the wastage, i. e., electricity from the oxygen of the air in the lungs; from the absorbents into the body from universal space; and from digestion of food; there is at once a lack of force sufficient for the excretory process to expel from the system the gross matter constantly accumulating, and also to circulate the blood equally.

Sec. 19. And it is the peculiar modification of electric force from such decaying, gross, morbid obstruction, which, added to the primary impression and acting on the nerves of sensation, is felt as disorder, and if the circulation of it is unequal, it is felt as pain or ache.

The peculiar diseased emanation or force from the decaying nerve of a tooth may often cause the adjoining teeth to ache, and the cheek may be so charged with it as to swell to twice its usual size.

The causes of disease, perhaps, cannot all be enumerated, but the principal or more evident ones may be noticed.

Depressing circumstances, or bad or good news, which are sometimes overpoweringly positive as a source of disorder and unequal circulation of life's forces to that degree that instead of a radiation from the vital parts outward, as in health, a reverse action takes place, and the blood recedes from the external surface inward, and congests the brain or other organs to that extent that loss of motion is the temporary or perhaps permanent result.

Extravagant exercise of the brain, or any or all of the organs of the body which attracts forces to them to that degree that a disordered circulation, congestion, or general exhaustion is induced.

Improper food : from which good blood is not made. Eating too much : which clogs and smothers a healthy action.

Eating not enough : the system is not sufficiently nourished.

Inhaling bad air, or drinking bad water: the blood is poisoned by them.

Intoxicating liquors, tobacco, narcotics, or poisons of any kind, whether called medicines or not : they cannot be appropriated as nourishment; they unhealthily excite the nervous system, and cause an unequal and disordered circulation ; they are pernicious in their tendencies; they destroy a healthy action according to the amount of their positive power.

SEC. 20. In fact, the causes of disorder may, wholly, or to a great extent, be stated as originating in the mind being too negative or passive, consequently receptive to any positive influence not in harmony with health ; not being sufficiently perceptive of the laws of health, and consequently not sufficiently positive to act in harmony with those laws ; the most positive idea, or impression, being for the time the ruling power.

Also, a proportion of the youth of both sexes being ignorant, and having many important lessons to learn, generally without wise teachers, and taking suggestions from their natural desires, which unwisely excites their generative organs, thus attracting life's forces to them more than their just due ; consequently other parts of the general system are robbed ; a symmetrical development is checked, and disease is perhaps slowly but surely the result ; the mind becomes weakened and the foundation is laid for nervous weakness, consumption, and the thousand miseries caused by a deficiency of force and an unequal circulation. Let all parents who have taken upon themselves the responsibility of rearing children cause them to read this, and to unmistakably understand whereabout that sunken rock is situated on which so many millions of both sexes have been wrecked in the morning of their existence without a word of warning.

SEC. 21. RULES APPLICABLE TO ALL CASES. Any cause that produces on the human system a more positive impression than previously existed, not in harmony with health, should be overcome or obliterated by a more positive impression by means in harmony with health.

In all cases of disease, whether acute or chronic, the vital forces are more or less reduced; consequently, waste matter is not sufficiently excreted, but remains as morbid obstruction to an equal and healthy circulation.

Therefore, to stimulate action sufficiently by means which is adapted and agrees with health, to overcome the morbid impression which previously existed, and at the same time to promote a discharge through all the clogged outlets for waste, thereby removing what obstructs an equal and healthy circulation, is a direct way to restore health.

HEALTH AND DISEASE DEFINED.

SEC. 22. A sufficiency of life's forces in the human body to digest wholesome food; for the blood-making process to proceed without hindrance; to circulate it equally, thereby nourishing every part of the body, and to excrete the waste through the natural outlets for it, constitutes health.

A deficiency of such forces, and the consequences, constitute disease.

The foregoing principles and general directions more particularly explained.

SEC. 23. The pressure of the atmosphere is about fifteen pounds to the square inch, and when life's forces are reduced in the body, there is not expansive power or warmth to sufficiently resist that pressure; consequently the fine channels or capillary vessels near the surface, and especially in the limbs and extremities, are contracted, which condition is not favorable to an equal circulation; also, the blood is colder and consequently thicker than when in health, which condition is often termed "taking cold."

When the system is not much disordered, mild means are often all that is necessary to restore lost action. But if life's forces are much reduced, let the patient be shielded from the full pressure of a dense atmosphere by wrapping in a blanket, or by warming the air in the room sufficiently by a fire; or, in some cases, by a steam-bath or a warm foot-bath; or, if in bed, by placing a hot stone wrapped in damp cloths to the feet, and if necessary, to the limbs and back; giving warm, stimulating drink, and induce action sufficient to perspire freely, not excepting the limbs and extremities.

If successful in determining the retained waste through the natural outlets for it sufficient to restore an equal circulation, all pain and ache and bad feeling is relieved; and if obstruction is all removed it is a permanent cure, by producing a stronger impression by means which are in harmony with health, than the disorderly impression which previously existed.

If life's forces have been reduced to that extent that the excretory process has been very much checked, the mucous surface of the stomach and intestines have become coated with a morbid accumulation of retained waste which will require time, more or less, to change and separate. The blood-making process has also been hindered by that morbid accumulation closing the mouths of absorbent vessels which in a healthy state open on that surface, and loss of strength is the result.

And at this point the inquiry is pertinent, What means will facilitate the separation and expulsion of that morbid accumulation? The answer is, that it has been proved by trials and practice that vegetable astringents combined with stimulants have a decided tendency to change and detach it. But when it becomes a loose mass in the stomach it is neither fit to digest or furnish electric force adapted to the needs of the physical body.

In such condition it sometimes causes nausea sufficient to expel it by vomiting; but, generally, assistance by lobelia is beneficial to produce that effect.

A stomach and system thus obstructed becomes in a short time much more difficult to cleanse than at the beginning of the disorder.

The kidneys are liable to become diseased from the morbid and cankered condition of the stomach; the liver is liable to become torpid, or inflamed from the bad quality of the blood; the mucous surfaces in the head become loaded with morbid accumulations, (catarrh) and sometimes causing deafness; the throat is often sore and cankered; the blood in its passage through the capillary vessels near the external surface is obstructed, and extra heat or fever is often the result.

A persistent use of stimulant, astringent, and perhaps relaxant medicine in quantity and frequency sufficient to produce action as near a healthy state as possible, in order to promote a due determination through all the outlets for waste; at the same time have the circulation equalized to the extremities, and as often as the stomach becomes loaded with a morbid accumulation, use lobelia sufficient to throw it off by vomiting, is a more powerful and efficient means to unclog the system than any plan by means of medicine which has yet been discovered, being also perfectly safe.

As soon as the morbid obstruction is sufficiently expelled from the organism, the use of stimulant and bitter medicine will restore and regulate the appetite until food will digest, furnish good blood, and restore the strength, when medicine is no longer necessary.

Sec. 24. The diseased magnetic influence from retained waste is often conducted by the nerves to a part distant from the principal seat of obstruction, and, as the part of the body where the greatest disturbance is felt has, according to the wisdom or foolishness of popular writers, determined the name of the disease, many mistakes have been made in consequence of it.

The decay of the nerve of a tooth which had been filled by a dentist, caused a very severe ache at the extremities of the temporal nerves at the side of the head without the sufferer suspecting that the decay of the tooth was the cause, as the filling kept the nerve from the oxygen of the air, and the tooth did not ache. When the tooth was taken away the ache went with it.

A morbid and inflamed condition of the stomach and bowels often cause headache ; and pain, ache, and weakness of the back, are often caused by disorder of the internal organs by means of the nerves of those organs which connect with the spine.

I have many times known of the mistake being made by popular physicians of pronouncing a disorder to be a disease of the spine, and the patient being subjected to torturing treatment, using caustic to make sores by burning in many places the length of the spine. without in the least benefiting them, but adding to their disorder. when the spine was really as sound as any part of them.   But the truth was, the internal organs were tuberculated and inflamed, causing great tenderness at the termination of the nerves from those organs to the spine, caused by a deficiency of life's forces, perspiration being checked in consequence, and the excretory process generally being sluggish.

I have also known of such patients who after being sentenced to die by allopathic physicians, to be cured by means not so popular as poisoning, but very effectual to unclog the whole system, and establish a natural, healthy, unobstructed condition.

Harmless vegetable medicines do not need to be administered with that great degree of caution and precision, as to quantity or frequency, which is indispensable in the use of mineral preparations or poisons.

SEC. 25.   A general rule by which to use medicine which agrees with a healthy state, is to give enough
4

to accomplish what is intended, bearing in mind to induce only the amount of action which is diffused and equalized.

But with poisons it is different, for if not enough is used to deaden sensibility, or cause death, in a degree, the struggle of life's forces against what obstructs the circulation is not overcome, and temporary relief is not obtained; and if too much is used, all sensibility is destroyed, and death or change is the result.

The means recommended in this work are simple, harmless, but powerful to unclog an obstructed state of the organism, and any one with sense enough to prepare nourishing food for use, can use them to restore lost action.

SEC. 26. It is not in the least necessary, indeed it is not for the best interest of community that any one should practice doctoring as a profession (as it is the pecuniary benefit of the doctor that the people are sick, and it would be a temptation to the unscrupulous to prescribe in accordance with the length of the purse of the patient).

It is essentially a plan of good nursing and assisting a healthy action, and the one who is capable of taking good care as nurse, should be able to know what means to use to restore health.

The foregoing remarks are not intended to apply to surgery as a profession. A critical and profound knowledge of anatomy and physiology, together with a natural love or adaptation for it, are indispensable qualifications of a good surgeon; and one good surgeon would be sufficient for a large extent of country if no more surgical operations were performed than what are absolutely necessary.

Sec. 27. I will briefly describe four methods by which a healthy action, when deficient, may be restored :

The first is, " Healing by laying on hands." An individual with an abundance of life's forces may benevolently direct them, through the hands, to one who is deficient, and in a negative or receptive condition ; and if the individual is magnetically sensitive, he or she, that is, the magnetizer, often consciously or unconsciously, has by influx or inspiration, assistance from the magnetism of others ; and if the conditions are favorable, and the magnetism is adapted to the condition of the patient, lost force may be restored, even by the law of equilibrium. This method has been practiced to some extent during many ages, sometimes by those who understood the philosophy of it, but, generally, by those who did not; and superstition and ignorance supplied its own explanation.

The second is by the positive, energetic mind of the patient overcoming a disorderly impression.

The third is by medicine, which will impart or induce magnetic or life force specially adapted to the emergency. By special adaptation, I mean a positive force which is evolved from substance in harmony with health, and competent to overcome the primary disorderly impression, from whatever cause ; and also, the diseased dynamic emanations from retained morbid obstruction, and to expel such obstruction from the organism through the natural outlets for waste matter.

The fourth is by electro-magnetism. The magnetic or galvanic currents evolved by the decomposition of metal in a battery, also the inductive currents, are not always well adapted to the condition of the nervous system; indeed, it is never so well adapted as human magnetism. But it is often beneficial if properly applied. From an experience of thirty-five years in the use of it, I can make this general statement: that when the patient has a fair amount of life's forces, but which circulate unequally, the currents can be so applied that an equal circulation can be promoted. But it cannot be equally relied upon when the patient is in a very weak or exhausted condition.

Good judgment in its use is very necessary, as it is capable of mischief if not rightly used.

For the information of those who possess batteries, with no more knowledge how to use them than to hold the conductors in their hands: and as many who make and sell them give a pamphlet of directions giving a wonderful account of the wonderful currents which they are capable of giving. I will give a simple direction how to test the currents of any battery. The primary current is magnetic, having a positive and negative pole, demonstrated by taking two platina wires for conductors, or attach a platina point to each of the copper conductors; then apply the platina points at a little distance apart in contact with white paper moistened with a solution of iodide of potash, and the paper under the point of the platina which conducts the positive current will be stained brown by the iodine being set free, while the other platina point will produce no discoloration of the paper. That proves a direct current, but with common batteries is of very little use, as it has not force enough to overcome the resistance of the physical body. The induced or secondary current is small in quantity but very intense, and by passing through the pole changer, what little polarity it possesses is reversed at every vibration of the spring: consequently, by testing it with the iodized paper both points of the platina will alternately be positive as the currents vibrate, and will discolor the paper. In most batteries with secondary currents the electric wave appears to be a little more in force in one conductor than the other, producing its effects a little more distinct in leaving a mark on the iodized paper, which entitles that conductor to be named comparatively the more positive. It is essential that the nature of the currents of a battery should be understood before using them. They should generally be applied to run in the direction of the ramification of the nerves.

In the fore part of this work I have indicated the location of the nerves from the spine communicating with the principal organs of the body. When by passing the fingers along the spine a tenderness is felt by pressure on any of the intervertebral spaces, ascertain where the nerve from that place leads to, as that organ is inflamed and probably tuberculated. It is often of benefit to apply the more positive pole of the induced current to the place on the spine thus indi-

4*

cated, while the other pole is passed opposite over the diseased organ. A mild current is generally better than a strong one, and only very mild currents, cautiously applied, should be used about the brain, heart, epigastrium, or any of the great nervous centers.

The design and limits of this work will not admit of more explanation. Battery magnetism cannot be even advertised into a permanent position as " a cure-all,"—indeed, the attempt would not often be made aside from commercial considerations, although in many cases it is a means of great value if reasonably applied.

FORMATION OF TUBERCLES, TUMORS, AND CANCERS.

SEC. 28. A particle of retained waste may obstruct a capillary vessel or a gland ; the vital force of the system being obstructed by it as a different conducting substance, it accumulates at that point, and an electric center or pole is formed, with attractive and contractive powers, around which lines of circulation other gross matter forms, and perhaps a wart or tumor is the result, which often at the commencement possesses but a feeble vitality, and which can sometimes be dissipated by the positive will of a second person placing his fingers in contact with it, as the forces of life follow to some extent the determination of the mind. If the magnetic force directed to it is of the same polarity, well adapted, and sufficiently potent, it repels and dissipates it. A cancer, boil, or tubercle commences in the same manner, and can sometimes be made, when forming, to disappear by the same means.

or by the current of a galvanic battery, if sufficiently well adapted and correctly applied : but as long as the cause or obstruction remains in the system the same liability exists for other formations.

Cancers, tumors, and tubercles have frequently been dissipated when forming by cleansing the system thoroughly by one or more lobelia emetics, aided by stimulants, astringents, and steam if necessary, to add warmth to the blood sufficient to circulate it equally, thus preventing the supply of morbid matter which feed such formations. If the circulation cannot be again established through obstructed places, they must of necessity suppurate. But in any case of disorder, whether hopeless or not, the patient is much more comfortable when the system is kept in as healthy state as is possible, that is, when the stomach and intestines are most free from an excess of morbid accumulation, so that the blood-making process and its equal circulation is not hindered.

SYNOPSIS OF GENERAL DIRECTIONS.

From every form of matter proceeds an emanation of force peculiar to that form, and as that form is decomposed, that peculiar quality is more free, and obeys the great law of love, affinity, or attraction ; and it is in the power of the spirit, according to its wisdom, to perceive and direct such forces for wise purposes.

Thus, when the human body is disordered and on the way to disorganization, the spirit has the power to perceive that life's forces by means of which it was kept in order is deficient in quantity and power, and needs assistance : perhaps it can perceive the direct causes of the trouble, perhaps not : but the fact remains evident.

SEC. 29. In all cases of disease the decaying waste of the system is more or less retained, and is throwing off its morbific influences which are conducted by the nerves to the most negative localities, producing its legitimate effects, as bad feeling, contraction or spasm. expansion or swelling, pain or ache. Now instead of searching for palliatives which are only dealing with effects (allopathic), or to overcome a lesser effect by a greater, which is not as permanent (homœopathic). a much wiser plan is to carefully assist life's forces to perform their usual task, to expel all obstruction, and equalize the circulation.

The use of stimulant and astringent medicine to cause more action ; to detach and determine the retained waste or the morbid muco-gelatinous coating from the mucous surfaces ; by the use of steam, if necessary, to warm and electrify the blood and induce a more active and equal circulation and perspiration : by the use of injections, if necessary, to cleanse the bowels and assist the circulation to the limbs and extremities ; and by the use of emetics, if necessary, to cleanse the stomach ; and then by the use of bitter medicine to regulate the appetite and the secretions of the liver, is an efficient and direct means to restore to a healthy condition. But a healthy condition is not attained until enough force is generated to establish an equal and natural perspiration by which the blood is cleansed.

FEVER.

SEC. 30. I have searched in vain the writings of many medical authors for a reasonable philosophy of fever and inflammation.

In passing an electric current through a copper wire of sufficient size, no remarkable effect is produced on the wire. it being a good conductor : but by cutting the wire. and inserting a piece of platina (a substance which is not as good a conductor), then by passing the current the platina becomes immediately heated. There seems to be a perfect analogy between the foregoing process and what takes place when life's forces have suddenly been conducted from the body, so that the excretive process has been checked. The capillary vessels have. in consequence, become clogged with retained waste which is not as good conducting matter as the living tissues ; the free circulating electricity of the system meets with resistance, and extra heat or fever is the result. The same effects from the same causes (obstruction) are observable in inflammation.

The direct cause of fevers, whatever they may be named, is the same : that is, deficiency or loss of force (from various more remote causes) to expel waste matter through the natural channels, the outlet through the skin being the greatest.

Thus what is called taking cold, and checking perspiration suddenly. is a common cause of febrile action.

If the use of stimulating and relaxing medicine, aided by steam if necessary, sufficient to induce a free and equal perspiration, is used soon enough to expel the retained waste, the extra heat goes off with it, and the fever is cured at once.

But if it is delayed until the stomach and intestines become coated with morbid matter so that the process of blood-making is hindered, it will require more time, and a persistent use of medicine (stimulant, astringent, and perhaps relaxant), which will have a tendency to mature and separate that morbid coating, or canker, and perhaps by an emetic to throw it off.

When the body is sufficiently cleansed, and the natural outlets for waste in active service, the blood-making process will proceed without hindrance, the circulation of it will be free and unobstructed, conveying nourishment to every part, and health and strength will be regained.

SEC. 31. When a certain locality becomes obstructed and consequently weak and negative, life's forces being positive to that locality are attracted, and the blood being circulated by these forces, is taken along by the current, and the congestion, swelling, heat, and inflammation is according to the amount of force, and resistance to the circulation of it.

Illustration.—A case of acute inflammation of the eyes. The patient had been treated by allopathic doctors, by leeches to the temples, and ice applied to cool the inflammation, low diet, etc. : in short, they had industriously applied what is named, regular treatment : that is, depletion to kill the heat, and they had killed so much of life's force, they had very nearly killed the patient.

Condition of the patient after the foregoing amount

of medical aid.—Cold hands and feet, and legs cold to the hips. Strong pressure of blood to the head; had been unable to sleep during three days and nights; intense pain through the pupil of both eyes; considered by the attending physicians as inevitably terminating in total blindness, if not loss of life.

Programme changed : doctors dismissed, another one called : water in teakettle heated as fast as possible ; two or three stones as large as double-fist put on the fire ; as soon as water was hot enough, took one-fourth teaspoonful of cayenne pepper, half teaspoonful of powdered bark of bayberry root, one-half teaspoonful of powdered lobelia leaves and pods, and added them to two-thirds of teacupful of hot water with sugar, and gave patient a tablespoonful of the tea. Then took away the ice from his head, and put a hot stone wrapped in damp cloths to his feet, laid two thicknesses of cotton cloth wet with water over his eyes, and put another hot stone to his hands ; then gave more of the medicine to keep the action of his stomach above that on the external surface.

The result was, the blood began to circulate to the extremities as his feet and hands became warm, which relieved his eyes from pain. The medicine promoted more action in his stomach, and increased the amount of forces which had been greatly reduced ; an equal circulation was established, thereby relieving the pressure to his head ; he was free from pain, and within ten minutes from the time he commenced to take the medicine, was asleep and in a gentle perspiration. So far, the work had to be done in a darkened room; the next day a little light was admitted and the eyes examined. They were of the color of raw beef; but by strictly attending to keeping the circulation of the blood equalized as in a healthy state, he was kept free from pain, his appetite and eyesight improved, and in one week he was out of doors.

In another similar case of inflammation of the eyes, it became necessary to thoroughly cleanse the stomach by emetic, and the bowels by injection, before the circulation of life's forces, and consequently the blood, could be equalized; the system of the patient being very much obstructed, being scrofulous.

Remember that in all cases, equalizing the circulation makes pain or ache an impossibility.

In mild cases of inflammation of the eyes, a poultice of slippery elm wet with a tea of red raspberry leaves, and applied, is often all that is necessary.

In all cases of disorder, drink a dose of more or less stimulating medicine before applications to the external surface, to be sure that the radiation of force shall proceed from the vitals outward.

HYDROPHOBIA.

SEC. 32. The individualized life of a dog whose organization of body is so much obstructed that the diseased magnetism or force from such obstruction stimulates it to bite (being mad and life cannot manifest except in accordance with the condition of the organism), and is successful in accomplishing its greatest desire, which is to bite, the saliva becomes a medium by which that diseased magnetism is conveyed to the one bitten ; and receiving a determinate impulse from the will of the dog, it becomes positive. Then begins the antagonism between the positive magnetism of the dog and the positive magnetism of the one bitten. If the one bitten can induce a more positive and determinate action by means which is more in harmony with a healthy condition than that induced by the dog, the greater overcomes and supersedes the less, and a cure is effected.

The above principles reduced to practice.—As soon as possible after being bitten take of African cayenne pepper half a teaspoonful, the bark of the root of bayberry pulverized, one teaspoonful, lobelia seed pulverized, one-third of a teaspoonful, one or two teaspoonfuls of sugar; to which add one tea-cup of hot water, and drink freely of it.

As soon as possible take a steam bath as thorough as it can be endured, so as to perspire freely ; then take a thorough lobelia emetic, keeping up a good action in the stomach in the meantime by cayenne and bayberry tea, and by the use of milk porridge or gruel, as it changes the fluids of the body and induces a strong determinate action from the vitals to the external surface and extremities, and through all the natural outlets for unhealthy accumulations. The general process for unclogging the system should be persevered in until it is thoroughly cleansed. In desperate cases relax the muscular system by lobelia sufficient to secure an equal circulation of the forces which circulate the blood, which secures the most favorable conditions for recovery.

5

SMALL-POX.

Sec. 33. This is a very contagious disease only to those who are in a morbid condition to receive it. The contagion or peculiar magnetic emanation may be taken in with the breath, but if the system is in perfect health the positive forces will soon expel it from the circulation.

Dr. Samuel Thomson says : "It is the highest state of canker and putrefaction that the human system is capable of receiving; the measles the next, and the canker-rash the third. Other disorders partake more or less of the same, which I am satisfied is a key to the whole; for by knowing how to cure this is a general rule to know how to cure all other cases, as the same means that will put out a large fire will extinguish a candle."

As the disease has caused so much anxiety perhaps it is well to notice its prominent peculiarities.

The symptoms generally are manifested as early as the tenth day, or from five to twenty days, after receiving the infection, which are pains in the head, back, loins, limbs, and generally nausea at the stomach, with cold chills, succeeded on the second or third day by considerable fever and sometimes pain and disturbance of the bowels, with inclination to vomit. The fever is most intense generally on the third or fourth day, just before the eruption makes its appearance, which is little sharp pimples on the face, neck, and breast. They continue to increase in size and number three or four days longer, when they appear on other parts of the body, but they often pass down in the course of twenty-four hours to the lower extremities.

When the pustules are not very numerous the feverish symptoms generally abate on the appearance of the eruptions. The pustules, if distinct, will generally be filled about the eighth or ninth day with a thick yellow matter that discharges. On the eleventh or twelfth day the eruptions begin to dry up, producing hardened scales which fall off, leaving the skin, temporarily, of a red or brown color.

But where the pustules are very numerous (owing to the system being previously in a morbid condition), on the eighth or ninth day the face is apt to swell and the eyelids to close, preceded by a hoarseness and difficulty of swallowing; the swelling of the face will generally abate about the twelfth day, or else there is considerable danger, which is seldom the case unless the patient has taken cold. If the fever does not subside at the breaking out of the eruption, but after the fifth or sixth day it increases and continues in a degree through the whole course of the disease, the disorder assumes a dangerous form, and it is in such cases only that it proves fatal, which is generally between the eighth and twelfth days from the commencement.

Small-pox is divided into two kinds—the distinct and confluent. In the first the eruptions are separate, and in the second they run together, forming large patches. The confluent is the most aggravated form. The eruption is frequently preceded by a diarrhœa, and the tongue covered by a dark or black coat, the saliva very thick and adhesive, and in the worst cases there seems to be a general tendency to putrescency.

But the fatality attending that disease has without doubt been more owing to the way it has been treated than to the disease. If, instead of the fashionable mode of treatment, which has been to physic, reduce the strength by low diet, and keeping them cold, a different or even contrary method was practiced, of cleansing the stomach and bowels and promoting a determination and discharge through the outlets for waste, it would be a direct way to restore to health; and those who have practiced it have proved it to be correct. By giving medicine more or less strong according to the action of the stomach, to promote a gentle perspiration, no difficulty need be experienced in treating small-pox. The mild stimulant and sudorific medicines, such as composition, pennyroyal, or catnip teas, would often be all that would be necessary. But if the system should be much clogged and in a very morbid state, it would be necessary to use means thorough enough to cleanse it. All that is necessary is to assist life's forces to expel the canker and putrefaction by keeping up a determination to the external surface. If the bowels do not move once in a day, use injections; but never in any case use purgatives, they would be liable to cause sudden death by the eruption receding inwards.

Sec. 34. I consider the practice of vaccination for kine-pox to be worse than foolish, for it inoculates and transmits all the diseased tendencies of the individual from whom the infection was taken, which makes the risk much greater than to have the small-pox. I have known many persons who have been very much diseased by that means. But I am certain it will be very earnestly recommended as long as it has a commercial value, and an interested class will still endeavor to have laws enacted to compel people to be vaccinated.

SUNSTROKE.

During the extreme heat of summer many people are prostrated, and even die, especially in large cities, from the effects of heat, generally from exposure to the direct rays of the sun. The philosophy of it is generally but little understood, as evidenced by the inefficient practices to restore them.

The sun's rays are magnetic and cause enough positive action or heat on the external surface to balance the positive forces generated internally, so that an equilibrium is produced, and instantly motion ceases. Life's forces being no longer generated in the usual manner to balance the pressure of the atmosphere, the capillary vessels of the limbs and extremities contract, and the blood is forced to the heart and brain. If that state is continued but a short time, as but one polarity is predominant, the fluids of the body stagnate, and decomposition must commence.

Means should be immediately used to give more power to the internal forces to change the equilibrium, so that the blood may circulate as before, and not press to the head.

I have found the most sure remedy in the worst cases, where decomposition had not already commenced, to be the third preparation of lobelia, one or two teaspoonsful taken into the stomach; then, if

5*

there is much external heat, conduct off enough by a cloth wet with water to produce a difference between the outward and inward heat ; when the proper difference is made the patient will breathe more free and easy.

But if life's forces have already been dissipated to that degree that the extremities are cold, by the blood receding from them, then apply cloths dipped and wrung from hot water to them, and by friction on the limbs induce, if possible, the blood to circulate in them again, which, if successful, will relieve the pressure to the head ; stimulating injections are also very efficient to assist the circulation downward to the limbs and extremities. If the means used to promote a natural circulation is successful, the patient will probably vomit a quantity of very viscous cold matter from the stomach, and will almost instantly feel more natural. In some cases, if decomposition had almost commenced, what is vomited will be very green or blue, and extremely offensive. The patient should then be treated as for ordinary disorder; that is, if necessary, cleanse the system sufficiently to establish a natural circulation and perspiration, and restore digestion.

But prevention is better than any quantity of cure. If it is perceived by a pressure to the head, or a slight dizziness, that the sun's rays are too powerful, immediate retirement to a shade, if possible, together with rest, may be sufficient, but if not, take a dose of pure stimulant, if obtainable, to increase the action of the stomach ; then, if necessary, conduct off enough of the external heat to promote a due distribution of the forces outward and inward.

Never use ice or ice-water in such cases. The use of ice in hot weather is an indirect and sometimes direct cause of many deaths every year.

Recollect that an equilibrium (of the forces which circulate the blood) outward and inward, is, if continued a short time, a cause of cessation of motion, and decomposition must ensue.

DROPSY.

A collection of watery fluid in the spongy texture between the skin and the muscles, or any of the cavities of the body, as the chest and abdomen.

It is caused by an exhaustion of life's forces generally, and as a consequence perspirable matter is retained in the system. It can be cured if attended to soon enough by cleansing the stomach thoroughly, and by causing enough action to expand the system sufficiently by warmth to establish a natural perspiration, and exhale the superabundance of watery matter, and then bitter and tonic medicine is good to regulate digestion.

RHEUMATISM.

With a great variety of symptoms, which are all caused by the circulation becoming obstructed and consequently unequal, may be cured by the means which will remove obstruction, and making conditions right for an equal circulation. Thousands have been cured, and can be again, by unclogging the system and establishing a natural perspiration. A mild attack of the complaint may often be cured by drinking warming medicine and bathing the affected part with stimulants more or less strong. Those who have taken mercury and other poisons (which should never be taken), are the most difficult to cure.

Human magnetism is sometimes efficacious to equalize the circulation of forces. Battery magnetism may be effectual if well adapted and applied correctly. but if the whole system is much obstructed it cannot be relied upon. Recollect that lobelia is a powerful means to unclog and equalize the circulation.

DIARRHŒA.

A thin, watery, slimy, or frothy evacuation of the bowels, caused directly by indigestible or irritating matter taken into the stomach, or by a loss of life's forces to that degree that perspiration is checked, and the fluids of the body recede from the external surface and determine inward to the mucous surfaces. In such condition the blood-making process is checked, causing great weakness. The bowels being especially weak, it is best to favor them by lying in a horizontal position most of the time, taking warming and astringent medicine to determine the forces from the vitals outward, and establish a natural perspiration, thus restoring a healthy action.

But if the effect cannot be easily produced, a more effectual way is to cleanse the stomach by emetic, and the bowels by injection, making use of steam, if necessary, to assist the determination to the external surface. During the disorder, and until there is no danger of a relapse. it is well to avoid solid food, as the digestive powers are not in working order, and it is apt to irritate. Milk porridge or gruel is better.

In mild cases of diarrhœa, a dose of the tea of bay-berry, with a teaspoonful of rheumatic drops and one or two teaspoonsful of sugar, drank warm, and a cloth wet with rheumatic drops applied to the bowels, is often all that is necessary. A dose of half a teaspoon-ful of black pepper, in milk, is sometimes well adapted to meet all requirements.

DYSENTERY.

is nearly the same as diarrhœa, and produced by nearly the same causes, except that blood is discharged in more or less quantity, with more acute inflamma-tion of the lower part of the bowels (technically the colon and rectum) ; in this condition also the blood-making process is checked, causing great weakness.

The direct means of cure is to remove the retained waste through the natural outlets, especially by per-spiration and by means which agree with health, thereby restoring healthy magnetic emanations.

The tincture of myrrh, having antiseptic qualities, or the rheumatic drops, is very good to add to drinks, and also to injections for all cases of looseness of the bowels, and especially when there is a tendency to putridity.

The forces from the astringents and diaphoretics, assisted, if necessary, by lobelia, to cleanse the stomach and equalize the circulation, are all the means necessary to use, except in a few instances when it may be of benefit to use steam in order to take off the pressure of the dense atmosphere, and thereby induce the blood to the external surface more effectually ; bearing it in mind that if the circulation of the blood is equalized, it does not press to the weak mucous surface of the bowels ; and if the fluids of the body are determined outward by perspiration, the frequent discharges by the bowels and griping will cease. Avoid solid food till the digestive powers are sufficiently recovered. In all cases of internal soreness or pain of the bowels, it may be well to bathe the external surface with rheumatic drops or liniment, or lay on a cloth wet with either of them, covered by a dry one to retain the warmth.

RECAPITULATION.

If life's forces are so much reduced that the fluids and waste of the body determines inward to the mucous surfaces, as in diarrhœa, dysentery, or cholera, if mild, warming, and astringent medicine is not sufficient to induce a natural and equal perspiration and to regulate, use means to thoroughly cleanse the whole system and work directly on the immediate cause, which is retained indigestible matter ; and as soon as perspiration is established equally, the discharges will cease. Keep warm, lie most of the time in a horizontal position, and avoid solid food till the stomach is strong enough to receive it without its causing irritation.

SCROFULA.

When an allopath is in a quandary what to name an aggregation of symptoms of a chronic disorder, the term scrofula is exceedingly convenient. Is it curable? Oh, yes! There are hundreds of sure cures for it, to be used externally, internally, and eternally. A vast number of compounds are continually advertised as specifics for scrofula ; but as they do not cure the disorder, somebody lies. The term scrofula, from *scrofa*, a sow, is representative of a mangy disease which once was supposed to be peculiar to swine. The symptoms are so numerous and variable that all of them never can be accurately recorded. But we may more reasonably state the cause of the disorder, which is simply exhaustion of life's forces, so that the waste of the system is not excreted sufficiently for health. An obstructed, morbid condition is the consequence. A healthy individual may become heated by unusual exercise so that perspiration is very profuse, then by cooling suddenly by bathing in cold water, or remaining in a current of cold air, or by any means by which the heat is suddenly conducted from the body, the liability is incurred of perspiration being permanently checked, eventually causing the great variety of symptoms consequent on general obstruction.

Instead of attempting to write a long list of symptoms attending a scrofulous condition, I will briefly state that a scrofulous patient does not excrete the waste matter of the blood, through the skin, in sufficient quantity. And means which will permanently establish a natural and equal perspiration will again establish health. But that means necessarily includes a general cleansing of the whole system, and living strictly according to the laws of health. Erysipelas, eczema, ecthyma, erythema, nettle-rash, rose-rash, watery pimples, herpes, rupia, lupus, mattery pimples, impetigo, scaly eruptions, leprosy, psoriasis, pityriasis, dry pimples, all simply represent an obstructed, morbid condition, which may be termed scrofulous. The use of mercury, antimony, lead, zinc, and preparations of minerals generally, improperly named medicines, is a prolific source of disorder, and a permanent cause of morbid action. The restoration to health of a patient who has taken mineral preparations is much more difficult, and sometimes impossible.

VENEREAL DISEASE.

Any cause by which the usual current of force from the internal to the external is reversed and determines to the mucous surface, promotes the secretion of an unusual amount of mucus which should be removed before it changes to a morbid or putrid condition. If it determines to the vagina, or neck of the uterus, by reason of induced weakness or irritation, it soon becames morbid to that degree that inflammation and ulceration is often the result, and by sexual intercourse that peculiar putrescency is transmitted to and absorbed by the male organ, causing in a few days an eruption of one or more pimples, being virtually an inoculation of morbid matter which is manifested as to quality according to the degree of putridity of the matter transmitted.

If it were not for the fact that some females have been very ignorant how to care for themselves in regard to cleanliness, and the deplorable consequences of carelessness, it would be a cause of the utmost astonishment that the disorder should ever have been allowed to make progress. It is at first simply a case of local canker, and can be met direct where it is with appropriate cleansing medicine; but after it is absorbed into the circulation and diffused in the whole system, then more thorough means should be used to expel it through the natural outlets for waste; remembering to keep up a good action of the stomach to promote a gentle perspiration. (Sec. 28.)

To the male who is so unfortunate as to absorb the putrescent secretion, I would say, you must be your own judge how honestly or wisely you have become diseased, but I will endeavor to direct you how to attain to a better condition. If an inflammation of the penis is observed, saturate a cloth with tincture of myrrh and thoroughly cleanse the surface; if it is not sufficient, apply a poultice of slippery-elm wet with warm water, to which add one or two teaspoonsful of tincture of myrrh, and drink freely of composition tea, and continue till the inflammation is cured. If any pimples form and fill with matter, saturate a small cloth with tincture of myrrh and rub them off sufficiently hard to cleanse the sore to the bottom, when they will commence to heal at once. If you have delayed cleansing yourself till the canker is taken into

the circulation, and if the urine is thick, or if a scalding sensation is felt while urinating, it will be necessary to use medicine to act on the kidneys and passages from them; for that purpose use the diuretic
compound as long as necessary, and perhaps it may
be necessary to cleanse the whole system, by the use
of means which has already been explained.

Very simple means are at first all that is necessary
to cleanse off local canker, but if delayed until it is
absorbed into the circulation, the whole system must
be cleansed from it by the use of means sufficiently
thorough to accomplish it (Secs. 21–29).

I have felt the imperative necessity of explaining
the nature and the means of cure of this disorder, as
it has during many ages been so miserably treated by
the use of mercury, which has been considered as a
specific. But the injurious effects of the so-called
medicine has been a thousand times worse than the
original disorder. It is difficult to select words sufficiently emphatic to convey a correct idea of the amount
of mischief which has been done by the use of it.

It is the morbid effects of mercury, added to the
syphilitic virus, which rots and destroys cartilage,
bones, and tissues; even when the original symptoms
are overcome by the greater change which mercury
produces, it remains in the system as disease, and if
the individual, male or female, becomes the parent of
offspring, the syphilitic and mercurial taint is transmitted to it, and, under the name of scrofula, to be
again cursed with poison called medicine, till the
grave hides the so-called *scientific* work.

Anciently, according to record, a spirit gave directions to the Jews to circumcise all the males. It has been supposed that the utility of the direction corresponded to the age in which it was given. But it is possible, in the present state of development of the reasoning faculties, to perceive that soap and water baths frequent enough for cleanliness, is preferable to mutilation.

It is also very significant that no direction was given to females, except at certain times, to separate themselves from the camp, as is practiced at the present day by the Indian women.

Perfect cleanliness, as a rule, should be sacredly observed by both male and female.

PULMONARY CONSUMPTION.

The predisposing causes of what is termed pulmonary consumption, are various (Sec. 18), but the direct cause is an obstructed condition of the organism and the negative condition of the lungs, by which tuberculous matter is deposited in them from the foul state of the blood. Although a common cold will often cause a morbid accumulation in the mucous surfaces of the head and bronchial tubes leading to the lungs, which may be thrown off by coughing and by the absorbents, yet the formation of tubercles, generally in the upper part of the right or left lung, or both, is the commencement of what has been termed the first stage of pulmonary consumption. Many persons have tubercles in their lungs during many years without their being aware of it, and even at last die of old age;

in such cases the forces of their system were enough to
excrete a sufficient amount of waste through the natural outlets for it, so that the blood was not in condition to cause those tubercles to enlarge and become so
hard as to stop the circulation through them, causing
ulceration, a process destructive to the air-cells or
substance of the lungs, which is termed the second
stage.   What is called the last stage, is when by
ulceration the lungs are very much destroyed, and the
whole system too much emaciated to recover.

The idea generally prevails that consumption is
incurable.   I admit that it is so by poisoning, a practice
which has prevailed during many ages.   I also admit
that the vital forces may be so far reduced and the
lungs destroyed by ulceration to that extent that progress in that direction may perhaps be hindered, but
cannot be entirely prevented.

I have, however, known of many being cured by the
means recommended in this work—to unclog the system generally, thus cleansing the blood and preventing the supply of morbid matter determining to the
lungs and preventing congestion of them by equalizing
the circulation—who have had their death-sentence
pronounced by those who have styled themselves "the
medical faculty."

The general treatment which consumptives have
received from said "faculty," has been according to
the "old *fogey*" plan of endeavoring to destroy symptoms and stop the struggle against what obstructed
the circulation, by deadening sensibility, and by palliatives which really reduced the life of the patient.

Invariably in a consumptive condition, as in all other disorderly states, the entire system is so much obstructed that the finer absorbent vessels concerned in the process of blood-making cannot for that purpose fully receive and elaborate the products of digestion, consequently the strength would fail, and although the appetite for food might be ravenous, its nutritious properties would not be appropriated.

Breathe good air, use wholesome food, and if when you take cold you generally have a cough, practice strengthening your lungs by deep breathing, so that the effects of colds will not determine there, and bid defiance to consumption, doctors who use poisons, and all pernicious influences.

FEVER AND AGUE—INTERMITTENT FEVER.

This disorder is produced by any means which reduces the forces of the system gradually to that degree that the first prominent struggle is an ague fit ; the blood recedes from the external surface, the teeth chatter, and a regular battle commences between cold and heat. The struggle continues generally till cold is partially overcome by the struggle, and is succeeded by a hot or feverish condition somewhat proportionate to the preceding chill, which is followed by sweating ; then gradually the system apparently regains its natural balance. After resting, perhaps one, two, or three days, the duration of the intervals depending on the obstructed condition, and also on the amount of vitality of the patient, another similar struggle is liable to occur.

6*

In this disorder, as in all others, generally, checked perspiration is the first result of a deficiency of force, and the means to regain lost action must be persistently used till that function is permanently established, and the natural amount of heat. or rather, force, which produces it is regained.

A gentleman came from one of the Western States to be cured, if possible, having not only thoroughly tried the allopathic drugs, viz., quinine, mercury, arsenic, etc., etc., etc., but also many advertised specifics, without being in the least benefited.

The mucous surface of his stomach and intestines was so coated with viscous matter that the absorbent vessels could not receive the products of digestion, even if the food was digested : consequently his flesh wasted away so that very little but skin and bones was left to shake. In general appearance he was much like an Egyptian mummy. dark, sallow, and extremely emaciated. When by premonitory symptoms a chill was expected, he was given a dose of cayenne and bayberry to warm and detach the mucus from the stomach ; also an injection of the same to warm his bowels and induce a better circulation in his limbs. Then he was put into a steam bath and the steam raised as hot as he could comfortably bear, at about the time when he would have had a chill if he had been out of it : but the cold being partially overcome. the shake was impossible. Then had him get into a bed and placed a hot stone wrapped in a damp cloth to his feet; then gave him a lobelia emetic which threw off about a quart of cold, viscous. morbid matter

from his stomach: then, after taking nourishment
and rest, gave him another light steam bath; the
warming and cleansing operation outside and inside
produced the first natural perspiration and natural
healthy feeling which he had felt for months; the
sweating threw out a very offensive, sticky, slimy
matter through the pores of the skin, which discolored
a towel to wipe it off, and changed the color of his
skin from a dead lead color to a florid or fresh com-
plexion. He was then rubbed over with the rheumatic
drops, and dressed. He was much elated with the
idea of being able to " out-general " the shakes. His
system was so much obstructed that the same treatment was continued at every recurrence of premonitory symptoms of a chill, for several days, until he
was free from obstruction and disorder.

I state the outlines of treatment in this case as it
was an extremely bad one. the worst I ever saw:
it had been made so principally by taking poisons
called " scientific medicinal preparations."

In ordinary, or the majorityof cases. less thorough
treatment may be effectual. The use of medicine to
stimulate the action of the stomach, as cayenne and
golden seal, or the composition powder. or a tea of
boneset with the addition of cayenne, or a tea of vervain, and if necessary an emetic to cleanse the
stomach, with a hot stone at the feet at night to warm
the blood and equalize the circulation of it, all or
each of the above means have been used with success.
The general action must be raised equal to a healthful
standard, so that waste matter is thrown off sufficiently.

But if the same causes, exhaustion by various means, malaria from stagnant decaying matter, damp cellars, etc., are continued, the same results will naturally follow. The conditions must be made favorable for recovery.

MEDICINES.

The means and medicines recommended in this work are selections from a large number as being the best and most efficacious of any yet discovered to promote a healthful state. There are a vast number of articles which sometimes would be good to use if the best could not be obtained, but as those which I have selected are common articles of commerce at the present day, and are all that are necessary to assist to a healthful condition, if the patient has enough vitality as a foundation to build upon, I judge it not best to add to the number, for a sufficiency of means, if well understood, are better than a great variety, with but an imperfect knowledge of them.

Vegetable medicines should be collected when they possess their greatest vigor, or at maturity before the frost injures them, when the leaves or seeds are desired.

Barks should be collected early in the spring when they peel off easily, or in the fall.

Roots should be dug when the sap is in them, either in the spring before the leaves grow or in autumn.

The leaves and seeds should be carefully selected free from dirt and parasites. Barks should be separ-

ated from the outer covering or ross, and the inner
bark only saved for use. Roots should be freed from
dirt and well cleansed, and if large, they should be
cut small to facilitate drying.

They should be dried in the shade, in a well-venti-
lated place, then pulverized, and kept protected from
the light and from dampness. Vegetable powders
have been kept a long time in a dry place in tin boxes.

EMETICS.

Lobelia inflata is a medicine of inestimable value
to temporarily relax the muscular system, and equalize the circulation of life's forces, and, consequently,
the blood. By its peculiar qualities it quickens the
nervous system to a keen sense. antagonistic to a
morbid influence. It is the most powerful and efficient
medicine yet discovered to equalize the circulation
and generally expel obstructive matter from the system.
Its nature is in harmony with a healthful state ; which
is demonstrated by the fact that when no morbid accumulation is in the system, no sickening effect is
produced by it. When a thorough means is necessary
to unload the stomach of an accumulation of morbid,
indigestible material, lobelia, if used in a proper
manner, will quicken sensibility so that the condition
of the stomach is felt as nauseous, and life's forces will
cause vomiting sufficient to throw off the morbid accumulation.

A distinctive characteristic of lobelia, and which
I believe is not possessed in as great degree by any
other medicine yet discovered, is that it will not cause

vomiting until sufficient action is induced to circulate the blood equally; thus being safe to use in all conditions, when necessary, as there is no injurious strain on any weak locality.

Lobelia has valuable properties as an emetic which are dissipated or distilled off by scalding; consequently scalding water should not be used in preparing it. It yields its properties to cold or warm water, wine, spirits, or vinegar.

There are conditions which are not favorable for lobelia to induce vomiting.

1. When the system is not warm and active enough; in which case make use of stimulants, more or less strong for the stomach, and it is generally best to add bayberry, witch-hazel, or red raspberry leaf tea, to act on the canker; then, if necessary, apply artificial heat by a steam bath, or by hot stones wrapped in damp cloths, to the feet and back to overpower the cold; bearing in mind to not apply heat to the external surface sufficient to balance the internal heat.

2. When the stomach is very sour: in which case dissolve nearly one-fourth of a level teaspoonful of bi-carbonate of soda in one-third teacup of warm water for a dose, which will have a tendency to neutralize the acid.

3. In a few instances it has been known to determine its forces through the pores of the skin by perspiration without vomiting; probably in consequence of the stomach being so much coated with tough adhesive mucus that medicine could not, for the time, produce sufficient action.

4. When no action can be raised in the stomach by medicine, or by artificial heat applied to the body, in short, when life's forces are so far reduced that not enough are left to build upon.

5. When there is no disease in the system; in which case it will have no effect otherwise than to cause a more rapid perspiration and a general feeling of warmth throughout the whole body.

The following directions for preparing lobelia for use are principally copied from Dr. Samuel Thomson's " Guide to Health : "

" This herb may be prepared for use in three different ways: 1. The powdered leaves and pods. 2. A tincture made from the green herb with spirit. 3. The seeds reduced to a fine powder and compounded with cayenne and myrrh with spirits.

" 1. Pound or grind the dried leaves and pods in a mortar to a fine powder; sift it through a fine sieve and preserve it from the air"—from light and from dampness. " This is the most common preparation, and may be given in many different ways, either by itself or compounded with other articles. For a common dose, take a teaspoonful of the powder, with the same quantity of sugar, in a teacup half full of warm water, or a tea of medicine for canker may be used instead of the water, to be taken all at one time or at three times at intervals of ten minutes. For a young child strain off the liquor, and give a part as circumstances shall require.

" 2. To prepare the tincture take the green herb in any stage of its growth," (perhaps the best time is

when it is in blossom), " pound them fine in a mortar
and add the same quantity of good spirit; work it
well together and press it hard through a fine cloth,
to get out all the juice; save the liquor in bottles
closely stoppered for use. Prepared in this manner it
is very convenient for internal or external use.

" For a dose take from a half to a teaspoonful; its
effects will be more certain if cayenne is added to it.

" 3. Reduce the seeds to a fine powder in a mortar,
and take half an ounce or a large spoonful of it, with
the same quantity of cayenne, and half teaspoonful of
powdered gum myrrh; add the whole to one gill of
good fourth-proof spirit, and keep in a close-stoppered
bottle for use. This preparation is intended for the
most violent attacks of disease, such as lock-jaw,
hydrophobia, fits, spasms, and in all cases of suspended
animation, where the vital spark is nearly extinct. It
will pervade the system like electricity, giving heat
and life to every part. In cases where the spasms are
very violent and the jaws become set, by pouring some
of this liquid into the mouth between the cheek and
teeth, as soon as it touches the glands of the roots of
the tongue the spasms will relax, and the jaws will
become loosened so that the mouth will open; then
give a dose of it, and as soon as the spasms have
abated, repeat it, and afterwards give medicine for
canker; this course 1 never knew to fail of giving
relief.

" It is good in less violent cases to bring out the
measles and small-pox. One of my agents cured a
man with it who had been bitten by a mad dog, and

I have not the least doubt of its being a specific for that disease. For a common dose take a teaspoonful.

" Much has been said of the power of the emetic herb, and some have expressed fears of it on that account : but I can assure the public that there is not the least danger in using it. I have given it to children a few hours old and persons of eighty years. It is most powerful in removing disease, but perfectly innocent in its effects.

" In regard to the quantity to be given as a dose, it is a matter of less consequence than is generally imagined. The most important thing is to give enough to produce the desired effect. If too little is given, it will worry the patient and do little good : if more is given than is necessary, the surplus will be thrown off, and is a waste of medicine. I have given directions what I consider as a proper dose in common cases, of the different preparations ; but still it must be left to the judgment of those who use it, how much to give : the best way will be to give the smallest prescribed dose first, and if not enough, give more. In all cases where the stomach is very cold and foul, its operation will be slow and uncertain : in which case give cayenne, which will assist it in doing its work.

" When this medicine is given to patients who have been diseased a long time, the symptoms indicating a crisis may not take place till they have been carried through from three to eight or more courses of medicine ; and the lower they have been, the more alarming will be the symptoms to those unacquainted with its effects. I have seen some lie and sob like a child

7

that had been punished, for two hours, not able to
speak or to raise their hands to their head, and the
next day be about and soon get well.

" In cases where they have taken considerable
opium, this medicine will in its operation produce the
same appearances and symptoms that are produced by
opium when first given, which having lain dormant is
roused by the enlivening qualities of the medicine.
and the patient will often be thrown into a senseless
state : the whole system will be one complete mass of
confusion, tumbling in every direction, and it may take
two or three to hold them on the bed ; sometimes they
grow cold, as though dying; remaining in this way
from two to eight hours, and then awake, like one
from sleep, after a good night's rest : being entirely
calm and sensible, as though nothing had ailed them.
It is seldom they ever have more than one of these
turns, and they generally begin to recover from that
time. I have been more particular in describing these
effects of the medicine, as they are very alarming to
those unacquainted with them, and in order to show
that there is no danger to be apprehended."

The foregoing " alarming symptons" may generally
be avoided by taking care to raise no more action in
the system by medicine than what the forces of life
can diffuse or equalize : always bearing it in mind to
assist an equal circulation, for on that depends success
in restoring health. In the earlier years of my practice I have observed what would have alarmed me if I
had not previously known of such occurrences, but
I never knew any bad results when they did occur.

I have given it to children and to the aged during a practice of over forty years, and I never knew it to act otherwise than as a friend to health, and as one of the most, if not the most valuable medicine with which I am acquainted to expel morbid accumulations, and to equalize the circulation of life's forces, and consequently the blood.

SEC. 35.   I have twice taken the third preparation of it at sea, when troubled with sea-sickness. It was in both cases an immediate relief, without vomiting; and I believe it to be a specific for that unpleasantness. especially if the patient is otherwise healthy.

Dr. Samuel Thomson has the honor of first making known the virtues of *Lobelia inflata* sufficiently to introduce it into extensive use.

His theory that " heat is life and cold is death," although not philosophically true, was a far better theory for practical purposes than any which had preceded it. And the medicines have not yet been discovered which are superior to those which he selected to unclog the physical system from obstruction, and because by those means he cured patients which allopathic practitioners by their drugs could not cure, no contemptible plan has been neglected by them* which promised the least success to bring him and the means which he recommended into disrepute; and the same infernal spirit. under the pretence of protecting the people, is active at the present day in endeavoring to influence legislatures to have laws

---

* I am glad to state that I believe there are some honorable exceptions among the craft.

enacted which virtually prohibit any one from curing the sick, under severe penalties, unless they qualify themselves at their school by learning their peculiar quackery and to give their favorite poisons. The dear people are fully able and intelligent enough to protect themselves, and honored be the legislators of Massachusetts who in the year 1877 had intelligence, honesty, and independence enough to object to that special deviltry.

BLUE VERVAIN (*Verbena hastata*) is bitter, nauseous, and slightly astringent. One or two teacupsful of the strong decoction will generally operate as an emetic. It is also good to promote perspiration, and will generally cure a fever in its first stages. Dr. Thomson ranks blue vervain next in value to lobelia as an emetic. After the stomach is thoroughly cleansed it does not generally produce nausea. It may be used as a tea made of the dry herb, or prepared in powder like lobelia.

In some sections of the country it is much used as a specific in fever and ague, and with good success, one dose having often effected a cure.

BONESET, or THOROUGHWORT (*Eupatorium perfoliatum*), has a disagreeable and bitter taste. yielding its medicinal qualities to water or alcohol.

A strong and warm infusion taken freely, generally operates as an emetic.

A weak tea drank warm and several times repeated, promotes perspiration.

An infusion of the flowers and leaves taken cold or made into a saturated solution with proof spirit, is very successfully used in intermittent fever or fever and ague, and by many is considered a specific for that complaint.

It is said that boiling decreases its emetic, and increases its laxative qualities.

STIMULANTS.

Pure stimulants are substances which excite or cause action in the human body by which in an obstructed condition a more active circulation is promoted and more warmth generated, without the danger of inflammation. Many articles are stimulating, and will excite action, but are so combined with other qualities that they are not pure stimulants, and perhaps should not be used. Stimulants which agree with a healthy condition cause a determination of force from the internal to the external surface, and promote perspiration. The best and most effective yet discovered is African cayenne pepper (*Capsicum baccatum*). The next best is that which grows in the West Indies.

The American red peppers (*Capsicum annuum*) are also a pure stimulant, but are not as strong as the African. Dose, from one-eighth to a teaspoonful in two-thirds of a teacupful of hot water sweetened.

The best cayenne is frequently adulterated with the inferior and cheaper kinds, and sometimes with meal, and paint or dye-stuffs, logwood, red saunders, ginger, red lead, and various articles to increase the quantity and weight. It is also, like all vegetable matter, liable to dampness and decay, which makes it unfit for

7*

use. I have generally obtained that which is good at wholesale stores in Boston and New York.

Other articles in which the stimulating quality predominates, but without acrid or irritating qualities, may be prudently used with success if the best cannot be obtained, such as pennyroyal, ginger, black-pepper, catnip, mayweed, peppermint, summer savory, yarrow, smartweed, spearmint.

SEC. 36. A pure stimulant held in the mouth will excite the glands to action, producing warmth without inflammation. It will also have a tendency to cleanse the mucous surface and leave the mouth clean and moist. If swallowed, it tends to produce the same effect in the stomach and intestines. The same effect would not be produced if combined with an acrid or irritating quality. The qualities of herbs can be to a great extent judged by that as a rule: that is, by chewing and holding in the mouth long enough for the juice to produce its determinate effects. If the mucous surface is left in a clean and healthy condition, it will produce the same effect (if swallowed) in the stomach and intestines, as it is the same surface continued.

All articles. stimulating, astringent, bitter, or any substance which has a definite quality, can often, by that rule, be judged as to the effect which would be produced by its use. Any substance which would not leave the mucous surface in that healthy condition, whether named food, drink, or medicine, should never be swallowed. (The foregoing rule was practiced by Dr. Thomson.)

Pure stimulants are generally well adapted to induce perspiration, but occasionally the very obstructed condition of the channels of circulation, by the resistance of such obstruction, generates more heat than is equally diffused. In such cases, relaxing the system gradually by small doses of lobelia, not in sufficient quantity to nauseate much, is beneficial : also, mild sudorifics, such at pennyroyal or sage teas, or whiteroot, may be used to promote perspiration, which, when induced, will carry off the extra heat. If perspiration cannot be induced by the reasonable use of such means, cleansing the stomach by lobelia, and the use of steam, if necessary, is a more efficient means. The use of medicine by injection is also often beneficial to cause a better circulation in the bowels and limbs, thus assisting an equal perspiration.

SEC. 37. Generally it is not best to produce action by stimulants or other means, more than is diffused or equalized, as pain and ache are caused in all cases by an unequal circulation. To illustrate,—the hands being exposed to the chilly winds of a very cold day, and the blood in the veins thickened by cold, so that the circulation in the capillary vessels is very feeble, a want to warm them is felt to be very imperative. If, then, the chilled hands are held near to a hot fire, a very uncomfortable ache or pain is the usual result when the hands are receiving warmth, until they are thoroughly warm and the circulation equalized.

The natural tendency of pure stimulants, when the influence which they impart is not prevented by too much obstruction, is to stimulate all the secretory vessels, the perspiratory function in particular, to greater activity, so that the extra heat induced by whatever obstructs is conducted off.

An individual sufficiently healthy may, in a hot day, or when the outward heat is very oppressive, drink pepper tea and take a steam bath sufficient to perspire freely, and feel a delicious sense of coolness as the result. On the same principle many persons have experienced the luxury of being cool in hot weather by drinking a cup of hot tea.

ASTRINGENTS.—*Medicines for canker.*—Vegetable astringents produce a contractive effect on the weakened or relaxed vessels of a mucous or serous surface to which they are applied. They should rarely be used to the extent of producing an uncomfortable feeling of contraction, and they should generally be combined more or less with a stimulant.

The bark of the root of BAYBERRY (*Myrica cerifera*) I place first on the list as being the best for general use. The usual quantity for a dose is one teaspoonful of the powdered bark of the root to a tea-cup two-thirds full of hot water, sweetened with sugar to suit the taste.

The powdered root of white POND-LILY (*Nymphæa odorata*) is also a very good astringent. It is excellent as a poultice for external application, being very cleansing for old sores. Dose, one teaspoonful of the powder in hot water, with sugar.

SUMAC (*Rhus glabrum*).—A tea of the bark, or

leaves, or berries. separate or combined, is also a good medicine to detach canker from the mucous surface. Dose, a teaspoonful of the pulverized bark or leaves to a teacup half full of hot water, with sugar.

WITCH-HAZEL (*Hamamelis Virginica*).—A tea of the leaves is very useful as an astringent medicine for canker, internal or external ; to be used freely.

RED RASPBERRY (*Rubus strigosus*).—A tea of the leaves of red raspberry is also a very good medicine for canker ; either separate or mixed with witch-hazel leaves. Either of them are more mildly astringent than the bayberry. pond-lily. or sumac, and is generally better adapted as medicine for children.

There are a great variety of herbs of an astringent quality, but I select the aforenamed as being the best, or all that is necessary for general use.

In some instances it may be beneficial to use more powerful astringents, as the root of cranesbill (*Geranium maculatum*), or the root of marsh rosemary (*Statice Caroliniana*), as outward applications. They are generally too astringent for internal use. They are sometimes used with benefit as applications for the piles.

RESTORATIVES.

BITTER AND TONIC MEDICINES.

They assist to promote the appetite and digestion, thus strengthening the body, when the stomach and intestines are sufficiently cleansed from morbid accumulations so that the process of blood-making is possible. The bitter and tonic quality is taken into the circulation and acts as a regulator of the secretions of the liver. The usual time of using bitter medicines is immediately before or after meals. Bitter medicines are beneficial after the whole system is sufficiently cleansed from obstruction, but they are of no benefit when taken into a foul stomach, often causing an uncomfortable sense of fullness in the head, or headache.

I select for use a few medicines only, which I consider to be the most useful.

GOLDEN-SEAL (*Hydrastis Canadensis*).—Golden-seal is an excellent bitter, tonic, and laxative. Its force will have a tendency to regulate the secretion of the liver, and restore the digestive powers.

General direction.—From a quarter of a teaspoonful to a teaspoonful of the powdered root, added to a teacup half full of hot water, with sugar, may be taken before or after each meal. It will generally give immediate relief when the food in a weak stomach causes distress.

BALMONY (*Clelone glabra*) is also an excellent regulator of the appetite. A common dose, a teaspoonful of the powdered leaves to a teacup two-thirds full of hot water, with sugar, before or after each meal.

POPLAR BARK (*Populus tremuloides*) is a very pleasant bitter, and can be used freely in a weak or debilitated condition to regulate the appetite. Its force has also a decided tendency to increase and regulate the secretion of urine, and to strengthen the kidneys. A strong decoction may be given with much benefit to children troubled with worms. Usual dose, one teaspoonful of the pulverized bark to a teacup half full of hot water, sweetened.

UNICORN (*Helonias dioicia*).—The root of the unicorn is an excellent bitter and tonic. Its force has also a peculiar tendency to strengthen the generative organs. It is very beneficial if used by women before and after child-birth for debility of the organs concerned in the process. It is also a good medicine for leucorrhœa. if the whole system generally is not so much obstructed as to prevent its specific action.

The usual dose is half a teaspoonful of the powdered root in a teacup half full of hot water, sweetened, two or three times per day. It is a very valuable medicine to prevent abortion and miscarriage: also useful in suppressed menstruation.

QUASSIA (*Picrœna excelsa*).—An intensely bitter tonic, sometimes used as a medicine to expel worms. An infusion of the rasped wood may be taken in quantity which is agreeable to the stomach, when a simple tonic impression is required. One drachm, or a heaped teaspoonful of the rasped wood to half a pint of water is about the right proportion for an infusion, of which from one to two tablespoonsful may be taken as a dose three or four times a day, or as it is agreeable to the stomach.

BARBERRY (*Berberis vulgaris*).—The bark of the stem and root of barberry is bitter, tonic, and laxative. It is useful to correct the bile and assist digestion. It

has been used with good effect in jaundice, and other
effects proceeding from a torpid liver. It is somewhat
similar in quality to golden-seal, but perhaps a little
more laxative, and may be used in the same way, or
according as it promotes a healthy action. A decoction of the berries is an agreeable acid drink, and also
useful as a gargle for a sore mouth and throat.

DIURETICS.

FIR BALSAM,—CANADA BALSAM (*Abies balsamea*).
—The fir balsam is a transparent fluid of nearly the
consistency of honey, which forms in blisters in the
bark of the tree. When fresh it is nearly colorless,
of fragrant odor, viscous, and to the taste slightly
bitter. When the stomach is disordered and the blood
becomes foul, not excreting enough of its waste by
perspiration, the kidneys often become disordered, the
urine thick or high-colored, often depositing a sedi-
ment, or in consequence of inflammation scant in
quantity. From one-fourth to half a teaspoonful of
the balsam dropped on sugar and taken once or twice
in a day, is an excellent remedy. It is very healing
in its nature, and is useful for weak or inflamed
mucous surfaces of the urethra, vagina, rectum, and
mucous surfaces generally. Combined with tincture
of myrrh it is an excellent remedial preparation for
leucorrhœa, gleets, gravelly complaints, and soreness
of the stomach and bowels.

QUEEN OF THE MEADOW (*Eupatoreum purpureum*).
—The root of queen of the meadow, called also gravel
root, is bitter, aromatic, astringent, and prominent as a

diuretic. It may be used with benefit in disorders of the urinary organs, such as gravel, high-colored, bloody, or turbid urine. It may also be used in dropsical affections to increase the secretion of urine when deficient in quantity. To prepare the infusion, steep an ounce of the bruised root in a pint of boiling water. A teacupful of this may be given every hour or two, or two or three times per day, as necessary.

PUMPKIN SEEDS (*Cucurbita pepo*).—An infusion of pumpkin seeds, made by bruising and adding one or two ounces to a pint of water, and drank freely, has been beneficial in cases of disorder of the urinary passages. It has sometimes been used with good success as a medicine to expel the tape-worm. An oil is obtained by expression of the seeds which has similar properties. Dose, six to twelve drops, several times a day.

CLEAVERS (*Galium aparine*).—An excellent diuretic. The herb is used in infusion, in scalding of the urine, inflammation of the kidneys and bladder, gravel, suppression of urine, and all obstructions of the urinary organs. An ounce and a half of the dried herb may be steeped in a pint of warm water and drank freely.

DIAPHORETICS.

WHITE ROOT, BUTTERFLY WEED, WIND ROOT, PLEURISY ROOT (*Asclepias tuberosa*).—This is said to be the only variety of *asclepias* that is destitute of a milky juice.

The root is a medicine of great value, and may be

8

used freely in a great variety of disorders to promote perspiration. When used very freely, it sometimes produces a mild, laxative effect on the bowels. It has been much used with good effect in inflammation of the pleura, hence, one of its names, pleurisy root. It is good to give to children with bowel complaints, and to expel wind. It has a tendency to produce perspiration without increasing the heat of the body. It can be freely used in fevers and eruptive diseases. Dose, one teaspoonful of the powdered root to a teacup two-thirds full of hot water, sweetened. It may be repeated several times per day if required.

CRAWLEY (*Corollorhiza odontorhiza*).—The root is the part used. It is excellent to promote perspiration, especially in low, feverish conditions. A common dose, half a teaspoonful of the powdered root in warm water can be given every half hour, or hour, until a gentle perspiration is induced.

Equal parts of crawley and pleurisy root make an excellent fever powder; they act without increasing the heat of the body.

PRICKLY ASH (*Zanthoxylum Americanum*).—A shrub growing in the Northern, Middle, and Western States.

The bark and berries are used for medicine, being stimulant, pungent, and diaphoretic. They are a good addition to other remedies for a cold, inactive, and sluggish condition of the system. If used very freely, it sometimes produces a pricking sensation of the skin. The berries, or seed and seed-vessels, have a fragrance resembling oil of lemons.

Tinctured in spirits, it has the reputation of curing epileptic fits. In some parts of the country it is much used for fever and ague. The usual dose of the powdered bark or seed-vessels is half a teaspoonful steeped in two-thirds of a teacupful of water, with sugar, and repeated two or three times a day.

A tincture may be prepared by adding two ounces of the powder to a pint of spirit.

A tree grows in the Southern States called prickly ash, which is not the *Americanum* or *fraxineum*, but the *tricarpium* of Elliot.

The southern prickly ash (*Aralia spinosa*) is very much stronger than that which grows in the more northern States.

SLIPPERY ELM (*Ulmus fulva*).—The inner bark of this tree abounds in mucilage, and is an excellent medicine for inflamed surfaces of the stomach, intestines, and urinary organs. The powder may be stirred in hot water (with or without sugar) in quantity sufficient to make a more or less thick mucilage. It is very nourishing, and may be used freely.

"A tablespoonful of the fine powder boiled in new milk is an excellent food for infants weaned from the breast; it prevents the bowel complaints to which they are liable."

Slippery elm bark, coarse ground, three parts, pounded cracker, one part, with warm water sufficient to form a poultice, makes one of the best applications ever discovered to allay inflammation, hasten the discharge of matter from tumors, and cleanse offensive sores. If necessary, ginger or cayenne may be added to make it more stimulating. Tincture of myrrh should be added if the poultice is applied to a part that has a tendency to putrescency.

MALE FERN (*Aspidium filix mas*).—A decoction of the root of male fern, in the proportion of an ounce of the dried root to a pint of water, and drank freely: or, two or three doses of ten drops each of the ethereal extract (oil of fern), at intervals of two or three hours, has been very effectual to destroy and expel the tape-worm.

SWEET FERN (*Comptonia asplenifolia*).—The plant is very common throughout the United States from New England to Carolina. It possesses the properties of the astringent and tonic balsams, and is useful in looseness of the bowels, and summer complaints of children. A decoction of the leaves, sweetened, is an excellent drink, producing a tonic effect, and also perspiration, without increasing the heat of the body. Many cases have been recorded of the tape-worm having been expelled by the use of it.

PENNYROYAL (*Hedeoma pulegioides*).—Pennyroyal is a mild stimulant, pleasantly pungent, and aromatic; when boiling water is poured on the dried herb, it is steeped sufficiently, and it is ready for use. The volatile properties are soon dissipated if left uncovered, or if the steeping process is continued. It is excellent as a mild, stimulating drink during the operation of an emetic, and as a stimulant for children.

SPEARMINT (*Mentha viridis*).—The herb is mildly stimulant, and may be freely used in sickness. Dr.

Thomson says,—"The most valuable property it possesses is to stop vomiting." An oil is obtained from it which possesses the medicinal properties of the herb.

PEPPERMINT (*Mentha piperita*).—This herb and its qualities are very generally well known. It may be used to promote perspiration, and is a good drink to relieve pain in the stomach and bowels of children. A few drops of the oil, or essence, dropped on sugar, is good to warm the stomach, but its effect is often only temporary.

MYRRH—THE GUM (*Balsamodendron myrrha*).—A tree growing in Arabia. yields a juice which hardens into a gum which is brought to this country and sold under the name of Turkey myrrh. When of good quality, it is, when broken, of a light brown color, pleasantly bitter, with a little pungency, slightly stimulating. aromatic. tonic, astringent, fragrant, and having antiseptic properties which make it very valuable as an addition to medicine for diarrhœa, dysentery, cholera, leucorrhœa, and all unhealthy determination to the mucous surfaces; also, in all cases where there is the least tendency to putrescency, and for spongy gums, sore mouth, offensive breath, and as an external application to weak backs, to wounds and sores (see Tincture).

SAGE (*Salvia officinalis*).—The tops and leaves of this common garden plant are slightly astringent, aromatic, and diaphoretic. The infusion freely used has sometimes prevented night sweats; it has also

8*

been used as a gargle in sore throat, and is sometimes relished by invalids as a drink.

It is somewhat relaxing to the muscular system.

CATNIP, or CATMINT (*Nepeta cataria*).—An infusion of the leaves drank warm promotes perspiration, and is often used with benefit in nervous irritability and febrile action. It has long been known by nurses as a good medicine for children in almost all their complaints. It has the reputation of being a bitter, pungent, aromatic, stimulant, and tonic. It has been successfully used in connection with the vapor bath, in suppression of the menses, and in curing colds and the disorders produced by them.

SPICY WINTERGREEN—PARTRIDGE-BERRY—TEA-BERRY —DEER-BERRY (*Gaultheria procumbens*).—The spicy wintergreen is similar as to quality and taste to the sweet-birch (*Betula lenta*); that is, a mild astringent and aromatic, with a very agreeable odor. It is chiefly used to impart an agreeable flavor to other medicines. A volatile oil is obtained from it by distillation, which possesses its peculiar flavor.

COMPOUNDS.

Tincture of myrrh.—To one quart of pure spirit, twenty-five per cent. above proof, add three and a half ounces of powdered gum myrrh. Shake the mixture thoroughly every day for one week.

Tincture of cayenne pepper.—To one quart of pure proof spirit add one ounce of cayenne pepper—to be shaken every day for one week, when it will be fit for use.

Mixtures for cough.—Molasses two parts, tincture of lobelia one part : add sufficient oil of pennyroyal to give it pungency.

Or, use honey instead of molasses. Some persons, however, cannot use honey, as it does not agree with the stomach.

Or, a jelly of Irish moss and lobelia, with sugar.

To be used at any time when the cough is troublesome, not in sufficient quantity to nauseate the stomach too much. The above preparations have the tendency to loosen the accumulations of morbid matter from the bronchial tubes, or the lungs, and enable the patient to raise it without much exhausting struggle.

While using means to loosen and easily raise the morbid accumulation which excites coughing, it is well to make conditions favorable for amendment, and also to use medicine to promote the excretion of waste through all other outlets, thereby diminishing the amount determining to the lungs or throat.

Rheumatic drops.—" One gallon fourth-proof spirits, one pound of powdered gum myrrh, one ounce of cayenne pepper.

" Put them into a stone jug and boil it a few minutes in a kettle of water, leaving the jug unstoppered. When settled, bottle it for use. It may be prepared without boiling by letting it stand in the jug five or six days, shaking it well every day."

This compound is to relieve pain and prevent mortification ; to be taken into the stomach, to be added to injections, or to be applied externally.

In almost every complaint it can be used to advantage ; from a half to a teaspoonful may be given alone, or the same quantity may be put into a dose of either of the medicines before mentioned, and may also be used to bathe with in all cases of external swellings or pains. It is a very convenient preparation to use when traveling, in case of pain in the bowels, or diarrhœa; by taking one or two teaspoonsful in the stomach and bathing the bowels, it is often a permanent relief.

Composition powder.—" Three pounds bayberry-root bark, one pound ginger, two ounces cayenne pepper, two ounces cloves. The whole to be pulverized, sifted, and mixed together.

"*Dose*—from one-fourth to a teaspoonful of the powder, and one teaspoonful of sugar to half a teacupful of boiling water, to be drank when sufficiently cool. It is intended to promote general action and warmth of the system and induce perspiration, consequently the conditions would be favorable by the patient being warm in bed, or by the fire covered by a blanket. It is very good to cure a cold in the first stages."

The two compounds—rheumatic drops, or hot drops, and composition powder—have become remedies of such extensive use in families, and also as there has been much miserable stuff made of worthless articles and sold under their names, I have concluded to give the original recipes, that those who have this work may prepare them from good articles, and not be defrauded. (They are as originally prepared by Dr. Thomson.)

Stimulating liniment (for outward application).—

To half a pint of tincture of cayenne, add one teaspoonful of sweet oil, two teaspoonsful of aqua-ammonia, one teaspoonful of oil of pennyroyal, one teaspoonful of glycerine.

Shake the mixture thoroughly before using. It is a very stimulating application, and sometimes, if used freely, the smarting is severe, but it does not produce irritation or inflammation. If the smarting is too severe, wet a cloth with water and wash it off, then bathe the part with milk, which will give relief. It is excellent for rheumatism.

Diuretic compound.—To one part of fur balsam, add four parts of rheumatic drops. As it does not all combine, it must be thoroughly shaken each time before using.

Usual dose, from half to a teaspoonful mixed with a teaspoonful of sugar.

One or two doses in the course of twenty-four hours is generally sufficient to regulate the secretion of urine, and as a remedy in inflammation of the kidneys or urethra.

It is very healing and antiseptic.

Cancer plaster of Dr. Thomson.—" Take the heads of red clover when in blossom (*Trifolium pratense*), fill a large kettle and boil them in water one hour: then take them out and fill the kettle with fresh ones, and boil them as before in the same liquor. Strain off and press the heads to get all the juice ; then simmer it over a slow fire till it is about the consistency of tar, when it will be fit for use. Be careful not to let it burn. When used it should be spread on a piece of bladder split and made soft. It is good to cure cancers and all old sores."

STEAM.

Steaming is very important as a means to assist, to increase, or restore the forces of life when very deficient, and in some cases is almost indispensable. A very simple and effective plan to give a steam bath is to heat three stones about the size of the double fist (the stones should not be too hard, as they will crack in the fire, and without an admixture of sulphur, as such are worthless for steaming, often causing the patient to feel faint); also have a teakettle of hot water ready; then pour the hot water into a tin pan until it is about half full; place the pan under a chair, and have the patient sit down in the chair, divested of all clothing; then wrap a large bed-quilt or blanket round both the chair and the patient, covering all up to the neck. Then take a hot stone in a tongs and lift up the blanket at one side sufficient to immerse the stone in the water enough to produce what steam is required. When the stone is too cool, exchange it for a hot one. It is best at first to have the steam nearly blood-warm, then raise it gradually in from five to ten minutes till it is very comfortable, or as the patient can bear it. The patient should, in the meantime, drink some cayenne or warm stimulating tea to keep up the action of the stomach; always bearing it in mind to keep the inward heat higher than the outward. If at any time the outward heat should accumulate so as to be equal to, or in equilibrium with, the heat in the lungs, the patient would be unable to breathe, and would faint; in which case a cloth wet with cold water, and passed across the forehead and stomach, would conduct off the outward heat and cause an immediate difference.

Sometimes it is best, and is very refreshing to the patient, when taking a steam bath, to occasionally pass a cloth wet with cold water across the face to lower the outward heat.

I was at one time called, in great haste, to attend a patient who had taken a steam bath, and was supposed to be dying, having fainted in the steam. I found him wrapped in a thick blanket and lying on a lounge, unable to breathe. By raising the blanket so as to let out the heated air, and passing a towel wet with cold water across his face and stomach, he could immediately breathe free, but he had absorbed more heat than he could retain, so it continually passed to the external surface more rapidly than the atmosphere could conduct it off: which necessitated conducting it off with a wet towel occasionally, for about ten minutes, before the natural balance was restored. Perhaps it is unnecessary to add that his cold was cured.

Some persons are in such an obstructed condition that they cannot bear the steam but a few minutes without feeling faint; occasionally, in such cases, by giving more cayenne or stimulating tea, and passing a cloth wet with cold water across the face and stomach, they will feel so much refreshed that they will finally bear a very good steam bath, with much benefit, from ten to twenty, thirty, or fifty minutes, with steam increased from blood-warm to one hundred or more degrees.

Steaming should not be measured by time or degrees of heat. The object to accomplish is to overcome the power of cold and its effects, and to promote a natural and equal perspiration and circulation.

It is sometimes advisable, when steamed enough to induce an equal circulation and perspiration, and if the patient is vigorous enough to bear it, to close the pores of the skin to a natural condition by a towel wet with water more or less cold. Some are vigorous enough to have cold water dashed over them, with benefit, to be followed immediately by a dry towel; if too much heat is conducted off, there will not be the necessary reaction or healthy glow immediately following it; in which case continue the steaming until a comfortable degree of warmth is produced; then rub over with a dry towel immediately, and get into bed or put on clothing, as health and strength permits.

The muscular system can be relaxed by warmth, as observable in the influence of a hot climate; and contracted by cold by suddenly conducting off too much of the caloric of the body. When a blacksmith desires to temper a piece of steel, he first makes it ductile by heat; then suddenly conducts off a sufficient amount of it to give the steel the desired temper, which it retains.

Many individuals have become scrofulous by cooling suddenly when very warm, so as to permanently check perspiration. Eating ice cream when the body is warm, sometimes produces the same results.

Instances are known of persons, when warm, suddenly conducting off so much heat by cold bathing, that they became cripples for life; although if they had relaxed themselves sufficiently by warming medicine, then taken a steam bath, and then cooled very gradually to the desired temperature, they could have been restored to a healthy state.

Steam bathing is very different from hot water baths. Steam not only cleanses the pores of the skin, but, as it condenses on the surface, it imparts its electricity to the capillary vessels, and induces an active and equal circulation through them, thus relieving local congestion.

By an adroit use of steam, and water more or less cold, a patient can be relaxed, or invigorated, and tempered as desired.

INJECTIONS.

When the physical system is so much exhausted as to prevent a natural excretion of waste through the skin by perspiration. the mucous surfaces secrete and exude more than when in health; the accumulation directly interferes with the digestion of food, clogging the absorbent vessels, consequently, the blood soon becomes foul, more or less of its waste being deposited in the liver, so that the bile is not properly prepared. Either constipation or looseness of the bowels is often the result.

If the stomach is much filled with the indigestible accumulation, an emetic will assist to throw it off. If the system is too sluggish to excrete through the bowels once in twenty-four hours, an injection will assist that action without producing any bad results, but will

9

really be beneficial to warm and induce a healthier condition generally. An injection for costiveness may be made of half a teaspoonful of fine bayberry, from one-fourth to a half teaspoonful of cayenne pepper, one-eighth of a teaspoonful of powdered lobelia, and one tablespoonful of molasses, or half teaspoonful of fine slippery elm, added to half a pint or one teacupful of hot water; use when cool enough, or a little warmer than blood-heat. Good judgment may sometimes vary the composition of them. They may be freely used when necessary. Better use a dozen when not needed, than omit one when it is needed.

If the patient is troubled with the piles, and much soreness, an injection composed of half teaspoonful of powdered bayberry, one teaspoonful of rheumatic drops, one-eighth of teaspoonful of powdered lobelia, and half teaspoonful of powdered slippery elm bark, would, perhaps, be as stimulating as could be comfortably used; it should also be strained from the dregs.

For producing a movement of the bowels in ordinary cases, an injection of a teaspoonful of the composition powder in a teacupful of hot water, and taken when cool enough, may be all that is necessary.

To cleanse and strengthen the bowels in case of diarrhœa, dysentery, cholera. or looseness of the bowels generally, injections are excellent. It is not necessary that they be made as stimulating and relaxing as those prepared for costiveness: they may be compounded of one teaspoonful of fine bayberry, and one teaspoonful of rheumatic drops : when cool enough to use, it should be poured off or strained from the dregs.

When the bowels are very weak and sore, as in dysentery, it is sometimes of benefit to add slippery elm and tincture of myrrh instead of the rheumatic drops.

Stimulating injections are very useful as a means to promote a more active circulation, and as the condition of the bowels governs the circulation through the limbs, it is an efficient means to warm cold feet.

Injections may be so compounded as to produce very many different effects on the system. They are excellent in cases of pressure to the head, as they tend to relieve congestion by equalizing the circulation. They are decidedly useful in all dangerous cases, and as they exert a powerful influence on all the neighboring organs, they are beneficial in promoting more action, as in suppressed menstruation, or inflammation of the bladder, uterus, kidneys, or bowels.

In one case of a patient who had been unable to swallow anything in five days in consequence of being choked with a substance lodged in the throat, the use of an injection composed of half a teaspoonful of bayberry, and one teaspoonful of pulverized lobelia seed, in half a teacupful of warm water, produced a vomiting effect which was an immediate relief.

By means of the absorbent vessels conveying their contents to the blood, nutritious injections have sometimes sustained life for a considerable length of time. One case is recorded of a patient who was unable to swallow being sustained sixty days by their use. They should consist of beef tea or some other nutritious fluid.

SEC. 38. And here I will notice the class of substances termed cathartic. The human family has been most outrageously cursed by their use. There is no substance which is used under that name but what acts by weakening the mucous surface of the stomach and bowels; and life by its forces struggles to expel it. Many substances by irritating determines the fluids of the body to the mucous membrane; a state which is contrary to a healthy condition. The natural radiation of force from the vitals outward, is reversed by its use, at the risk of establishing a disordered action generally, and there is no hazard in the statement that of a certain number of patients who take purgatives, the one who takes the most will be the first to die, although some have enough vitality to endure abuse a long time.

SEC. 39. There is a vast difference between irritation and stimulation. Irritation will weaken and cause inflammation, while pure stimulants are powerful to cure inflammation. In their effects they are opposites. Cayenne pepper will often by stimulating the stomach, liver, and intestines to greater activity, cause a movement of the bowels and remedy costiveness. Golden-seal, white-root, or boneset will often act similarly, but they do not, like purgatives, weaken or inflame the mucous membrane, or check a natural perspiration.

SPECIFIC ACTION OF MEDICINES.

There are medicines which possess a certain degree of affinity for certain organs of the body sufficient to determine to such organs, and either, according to their nature, promote or hinder their healthy functions : but the stomach or the whole system may be so much obstructed that the positive morbid influence from such decaying obstruction may counteract the natural tendencies of such medicine and render it of no decisive effect, according to the principle that the greater overcomes the less. In such cases a general process of unclogging is necessary ; after the stomach is sufficiently cleansed and the circulation of the blood equalized and active, the specific qualities of medicine may be confidently relied upon. But, in most cases, a general cleansing from obstruction sufficient to establish a natural and equal perspiration supersedes the necessity of more specific action.

When the forces of the system are so far reduced that a natural and equal perspiration is hindered, lost action may often be restored by the use of warming teas, as cayenne or pennyroyal : or, if the excretive process has been very much checked, perhaps it may be better to take a vapor bath, and an emetic to cleanse the stomach, as it is a very effectual process for that purpose. When the stomach is not very foul, simple means, as doses of white-root, crawley, thoroughwort, or small doses of lobelia to relax and warm the system, are often all that is necessary.

When from the same causes the excretion through the kidneys is checked, if the stomach is not too foul, that function can be promoted by the use of a tea of poplar bark, or queen of the meadow, clivers, pumpkin seeds, or balsam of fir. But if the excretion through the kidneys is too profuse, it is in consequence

9*

of perspiration through the pores of the skin being checked, which if promoted relieves the kidneys.

SEC. 40. A constipated state of the bowels is caused by perspiration being checked, so that the blood becomes foul, and consequently the liver becomes obstructed and torpid, and does not secrete a sufficiency of bile for the requirements of the digestive process. The use of food which is easily digested, together with the use of cayenne to stimulate, or the use of goldenseal, or white-root, together with the use of injections if necessary, will often be sufficient to regulate. If not, cleanse the stomach, and persist in the use of means to rouse up the system to more action, so that the blood may be sufficiently cleansed, and the liver will be enabled to perform its duty. More time and patience is necessary to regulate a disordered action of the liver on account of the naturally slow circulation through that organ. The liver may be so much obstructed that the bile may be very acrid, causing inflammation not only of the liver but the duodenum, or second stomach, and intestines, causing common and sick-headaches; and in that condition, if any indigestible article of food is taken and acts as an irritation, perspirable matter is very liable to determine inward to the mucous surface, and diarrhœa, dysentery, or cholera is a possible result.

Constipation can generally be relieved by changing the diet. Only articles of food which are easily digested, the unbolted-flour bread, vegetables, tapioca jelly, wheat jelly, and many other articles, avoiding

the use of coffee, tea, tobacco, ardent spirits, and
everything which is perceived as injurious.

SEC. 41. In consequence of torpidity of the liver
the bile is so thick and poorly prepared that sometimes
an electric center forms as a nucleus around which
matter accumulates and forms stones from the size of
small grains to that of a walnut. They are found
in the gall bladder, and sometimes in the cystic and
hepatic ducts, which in passing causes great pain. In
such cases, relaxing the stomach and whole system
thoroughly by a lobelia emetic, is a very good means
to enable the stones to pass into the bowels, when the
pain will cease.

REPRODUCTIVE ORGANS—DISORDERS AND REMEDIES.

The principal reproductive organs contained in the
female pelvis is the uterus and its appendages, the
fallopian tubes, the ovaries, and the vagina or canal
which leads from the uterus to the external parts
of generation ; all of which are plentifully supplied
with vessels of circulation and with sensitive and motor
nerves. Being organs which sometimes appropriate
a large share of the vital forces of the system, it is
very important that females should understand thoroughly how to care for themselves so as to attain and
maintain the highest degree of health ; but of such
knowledge they have generally been most deplorably
ignorant. Depending on doctors to think for them,
they for ages have been the subjects of experiments
with poisons, directed in accordance with false and
pernicious theories, and it is principally from that

cause that scarcely a healthy woman exists in the
civilized parts of the earth at the present day; and
as unhealthy parents cannot produce healthy offspring, much of disease is inherited.

The female until the age of puberty, which in this
climate is about the age of fourteen, is subject in common with the male to become directly obstructed, by
loss of force from whatever cause, sufficient to check
the excretive process more or less through the skin,
the lungs, the bowels, and the kidneys.

But at the age of puberty her system undergoes a
change more or less perfect according to the amount
of vital force which strongly determines to the
reproductive organs, taking along in its currents the
results of unusual activity of the secretive organs, and
passes it through the vessels which open on the internal surface of the uterus, from whence it is expelled,
as unappropriated matter, by the menstrual discharge.

In full health and vigor such discharge is recurrent
every month, and usually continues three or four days;
in quantity, five or six ounces. It very rarely makes
its appearance during gestation or during the time of
nursing, and conception does not take place until after
menstruation.

If sufficiently healthy and vigorous, the monthly
discharges, unless interrupted by pregnancy, continues
until nearly the age of fifty, when another important
change takes place, and the other outlets for waste
perform all the excretive requirements of the system
as before.

The correct standard for an amount of vital force

for health, in all cases, for young or old, male or female, is an amount sufficient to eliminate the worn-out or waste matter of the system through the natural outlets appropriate for the same.

Illustration.—If life's forces are deficient, so that perspirable matter is checked in its passage through the skin, it may be retained, and named dropsy ; or, determined to the mucous surface of the intestines, as diarrhœa ; or, through the kidneys, as diabetes, in male or female ; and none of the afore-mentioned cases of disorder can be permanently cured except by restoring enough force to again restore the determination of a due proportion of waste through the skin as perspiration.

But the consequences of a lack of life's force sufficient to continue a healthy excretory process in the female, especially at the commencement of or during menstruation, is the cause of innumerable difficulties, which may be remedied, at first, by very simple and harmless means.

I believe I can truthfully say, that during a practice of forty years, I have heard the statement of thousands of females who have dated their poor health from the time when they took cold at some time during the menstrual period, and that function was more or less checked. Many of them, when young, and without any knowledge of the natural changes in their system, and having mothers who were so excessively and morbidly modest that they had not informed their daughters about what intimately concerned their life and health ; being, also, almost as ignorant of the subject as their children. and their doctor's prescriptions not being composed of substances in harmony with health, were worse than inefficient, as they added to the already morbid condition of the patient.

The result of a deficiency of force to eliminate the proper proportion of waste through the external surface, is an undue determination to the mucous surface, sometimes more in the head than elsewhere. causing catarrh.

A direct way to remedy the difficulty before the decaying accumulation in the head becomes morbid and corrosive, is to promote action sufficient to again determine to the external surface, by perspiration. so as to stop the accumulation in the head, and by the use of fine bayberry as snuff, to sneeze out the accumulation already there.

Sometimes the result of the same cause determines to the stomach and bowels. The use of simple stimulating or sweating medicine will restore lost action. if the stomach is not too much burdened by the accumulation ; in which case, throw it off by an emetic and cleanse the bowels by an injection, so as to establish a natural and equal perspiration and circulation. and lost force is regained.

SEC. 42. Sometimes. also. from the same cause, the determination is to the vagina, causing leucorrhœa. As in the above cases. restore the lost force of the system sufficiently to determine the fluids to the external surface by perspiration. and at the same time apply medicine directly by injection, composed of the mild astringents, as red-raspberry or witch-hazel leaf tea, or the stronger, as bayberry or white pond-lily root steeped in hot water; when cooled to nearly blood heat strain off nearly a teacupful of the tea, and add one teaspoonful, or more, of the tincture of myrrh. The above formula for a female injection can be changed as the case requires. The myrrh can often be used freely with good results as a direct means to cleanse and strengthen.

As it is with the mucous surfaces in the cavities of the head, when by reason of excess of secretion of mucus it remains and becomes morbid, being in a state of decay, causing what is termed catarrh, so it is with the mucous surface of the stomach and intestines, the secretion soon changing to canker, and, if it remains long enough, to putrefaction. It is the same when determined to the mucous surface of the vagina, and should be met with medicine which will both cleanse and strengthen that surface, as it is always, in such cases, weak and relaxed.

If the original cause is continued, it often causes in addition, tuberculation, and, ultimately, ulceration of the vagina and neck of the uterus, with inflammation, a state of disorder which needs, in addition to a thorough cleansing of the system, local applications to cleanse and strengthen, persistently applied. Recollect that equalizing the circulation relieves congestion in any part of the body.

Owing to retained waste, which soon becomes morbid in the system, a canker or virus is sometimes determined to the external surface of the genital organs, causing an extremely irritating, itching sensation. Rubbing the surface with a cloth wet with the tincture of myrrh, will almost invariably remedy the local difficulty. In some cases it may be necessary to attend to the causes of it.

Now, I propose a *revolution*, of greater import than the French or the American : That every female carefully reads this work, harmonizes her mind with the inevitable laws of progress, learns how to doctor herself, and attain to the highest degree of health of which her organism is capable.

REPRODUCTION.

When this work was commenced, it was not intended to notice extensively the exceedingly various phases of disorderly action, but, rather, to explain principles necessary for establishing health ; being cautious of the miserable mistake of the allopaths in giving undue prominence to ever-varying symptoms, or effects, instead of sufficiently studying causes.

The allopaths' hobby of constantly observing symptoms or effects of disorderly action of force by obstruction, and giving names to an aggregation of symptoms as a distinct disease, and then to try to obliterate the symptoms by a more powerful impression by deleterious means, and not at all adapted to promote a sufficient amount of force, the deficiency of which was the origin of an obstructed state which caused the bad symptoms, has made an abstruse subject doubly mysterious. In fact, the causes of health and disease has not generally been enough thought of to elevate the mind on that subject up to the level of common sense.

The reproduction of the human species is a work for which the female organism is in every respect perfectly adapted; and, if the laws of health were strictly obeyed, there would be but slight disturbance of a general healthy circulation.

After impregnation of the ovum, or egg, a portion of life's forces are directed somewhat out of their usual course, to continue and perfect the process of gestation; and, in due time, or about nine months, to introduce into this rudimental school a form composed of organs well adapted to represent the germs of faculties for further development and individualization of spirit.

It is, indeed, seldom that pregnancy does not, more or less, disturb the digestive function, from the reason that in the parts of the earth that are termed highly civilized, very many causes have a tendency to reduce physical vigor far below a healthy condition; consequently, the action of the stomach being feeble, and the reproductive system being more active, a radiation of force proceeds from them to the stomach and other parts, causing morning sickness and other disturbances; whereas, in an orderly condition, a radiation of force should proceed from the stomach and vital organs to all other parts, they being, in a state of health and vigor, the most active.

SEC. 43. And here I will state it as a rule, that from the point or locality where the greatest activity, or change, prevails, a positive force radiates to other more negative localities.

Illustration.—If pain moves from any part of the

body towards the vital organs, sufficiently stimulating medicine, adding lobelia, if necessary, will drive it the other way.

For disorders during gestation, the same means may be used to promote a healthy state and to regulate, as at any other time ; *i. e.*, if the forces of the system are reduced below the amount which is necessary for the blood to throw off its waste by perspiration, stimulate to greater activity, and equalize the action by the use of steam-baths and emetics, if necessary, and by the use of injections, if necessary, to warm the bowels and induce the blood to circulate sufficiently in the limbs and feet to warm them.

It is very important that near the time of delivery a woman should be in possession of all her physical powers ; she should be in a condition that a dose of moderately-stimulating medicine would produce a perspiration over the whole body, the limbs not excepted. Recollect that when the system is sufficiently warm, that the blood circulates in the finer channels, where it cannot when in a colder state, and that when the forces that circulate the blood are equalized there is no pressure to any weak part, and pain and ache are prevented ; and that were the system sufficiently relaxed and warm, the necessary distension during labor is more easily accomplished ; and also, that lobelia is the safest and best relaxing medicine yet discovered.

Pregnancy should by no means be considered as a diseased or disorderly condition ; but, as many changes occur during gestation, it is very important that a healthy condition should be maintained, so that the vital forces may be fully competent to circulate the blood equally, giving nutriment and strength to all parts. Sufficient directions have been already given to promote action when deficient.

The most prominent difficulty in parturition arises from the system being in such a cold, inactive state that relaxation and an equal circulation is not possible without more or less struggle, which is weakening.

If reasonable means are used to promote a warming and relaxing effect sufficient to equalize the circulation, the strength is retained.

If proper means are used to cleanse the system sufficiently to equalize the circulation of the blood, and relax rigidity of the muscles by warmth, the use of force from those who assist would not be necessary; very much mischief has been caused by substituting art for nature.

As an illustration, I will state the case of a woman who had been three days in labor without much progress. The doctress in attendance had, previously, fair success as a midwife and nurse, but lacked the most important knowledge—how to equalize the circulation, and relax the muscular system, by warmth and medicine. She was so extremely anxious that everything should be right that she had made frequent examinations, which had worried the patient and checked the natural process of dilatation. They finally became alarmed, and concluded to send for another to take charge. The patient was found to be in rather a cold and inactive condition, having been without sleep, tired. and somewhat discouraged, and, being somewhat advanced in years, anticipating a hard labor.

Preparations for treatment were commenced by adding to a teacup two-thirds full of hot water, half a teaspoonful of fine bayberry bark, of the root, and half a teaspoonful of cayenne pepper : when cool enough to take, poured it from the dregs. and added one teaspoonful of pulverized lobelia-seed and two teaspoonsful of sugar, and gave her a tablespoonful of the tea ; then placed a hot stone wrapped in a damp cloth to her feet : then gave her the rest of the dose of medicine. The whole system soon began to relax, by warmth, and, as several women were present, all but one were requested to leave the room and give the patient an opportunity to rest. She soon was asleep and slept about two hours. When she awoke she was very much refreshed, and in a fine perspiration, which was equal on her limbs as well as body. The labor pains soon commenced. and in about two hours and a half from the time she took the medicine a fine boy was added to the population, and by giving medicine to sustain the action and equal circulation, the placenta, after about four hours' rest, was expelled and all after pains prevented.

In every neighborhood there are women who have capacity and good judgment sufficient to qualify themselves for acting as midwives and nurses ; and thus restore the business to those to whom it naturally belongs. If it was properly conducted, invariably without any force being used, the necessity for a surgeon would never occur. except, perhaps, in very rare cases of malformation, as in case of dwarfs.

Every woman who by knowledge has obtained wisdom sufficiently, knows that it is at the risk of health and perhaps life that abortion, or miscarriage, is produced. When, by any cause, there is a tendency or danger of either event, equalizing the circulation of the blood is the most powerful means yet discovered to prevent it; it will also regulate untimely pains in parturition.

There are so few healthy women at the present day that it is almost a necessity that a course of cleansing from obstruction, by the use of one or more lobelia emetics, and steam bathing, if necessary, to add warmth to the blood and to open the pores of the skin, should be used from two or three weeks to a few days before confinement, to insure the best and easiest results. for the more equal the blood circulates the less liability for pain and hemorrhage.

MISCELLANEOUS REMARKS.

Action induced by medicine more than is diffused, is capable of producing morbid effects according to the unequal circulation induced. When it is evident that morbid effects are thus produced, and not by the natural progress of the disease, the medicine should be discontinued or changed, and it is generally best if a patient is evidently gaining in health, although slowly, to hesitate to materially change the treatment.

Sec. 44. No specific quantities of medicine can be wisely prescribed for children. All the medicines recommended in this work are harmless if prudently used, and the doses can be regulated in amount and

10*

frequency according to their age, condition, and the amount of vitality which they possess to equalize the action produced by it. The best judgment, combined with a natural love for children, constitutes the best qualities for a nurse to assist them to a healthy state.

The most prominent disorders of children arise from a deficiency of force to excrete the waste matter through the natural outlets for it, consequently the stomach becomes clogged with a superabundance of mucus, which is often by natural effort thrown off by vomiting.

When life's forces are not sufficient to soon regulate to a healthy standard, assistance can be given them by warming the stomach : perhaps pennyroyal tea will be enough stimulating, or cayenne tea with sugar and milk ; if the stomach has become morbid or cankered, add red raspberry or witch-hazel leaf tea, or if necessary, the stronger astringent, as bayberry. If the bowels are disordered, it is sometimes necessary to use the same medicine direct by injection ; the tincture of myrrh should be added if there is any tendency to putrescency. If wind accumulates in the stomach or bowels, a tea of the white-root will be useful. If the child is restless or sleepless, a few drops of lobelia, or enough to equalize the circulation, will quiet, or give enough to vomit, if necessary. A tea of the poplar bark is often excellent for children who are inclined to be troubled with worms. As the bitter quality of it assists digestion, worms will never trouble a child with good digestive powers. The material in which worms breed is a superabundance of mucus.

I have never yet failed to relieve all grades of throat disorders, croup. diphtheria, etc., by a thorough lobelia emetic when necessary. Invariably there is canker in the stomach in all throat disorders. In mild cases a gargle of any of the stimulant and astringent medicines is useful, and is often all that is necessary.

The important fact should never be forgotten that in all cases of obstruction of the organism, life, by means of its forces, struggles to throw off what obstructs : and no individualized being has enough wisdom to know certainly how much interference will assist or hinder. The amount of good judgment depends on how much is perceived of truth. If those only of the number of cases which are absolutely known to have been cured by medicine were recorded, they would amount to only a small pamphlet. It is not a demonstration that because certain means were used and the patient recovered, that the cure was the consequence of the means used : because other agencies were at work ; the means relied on may have even retarded the recovery. Consequently, how preposterous and worse than foolish is the practice of using poisonous preparations, well knowing they are liable to produce very different effects from what is intended, and it is extremely absurd to name such practice scientific.

Sec. 45. Every individual is endowed with a certain amount of vitality which with favorable conditions will last a certain length of time. But while one may not have enough to last through infancy or youth, another may live one hundred or more years. All the medicine in the world cannot add to the natural

amount which is possessed : it can only assist to make
the conditions favorable for the natural stock of vitality on hand to operate to the extent of its limits.   It
has been observed that, excepting accidents and special
causes, the limits of an individual is about the medium
time between the ages of the parents.   But there are
exceptions to general laws.

I believe the time, however distant, may be reasonably expected when by natural progress a less number
of children will be born, but of far better quality,
with much greater physical and mental endowments.

A vast number of individuals at the present day,
generally in consequence of the pernicious influence
of drugs, especially the various preparations of mercury, never have a natural perspiration : or if they
perspire at all, it is about the neck or forehead. but
not on the limbs.   Many are born with that diseased
condition in consequence of the disordered condition
of the parents.   Generally, if by unusual exertion the
heat of the body is increased, the skin grows very
uncomfortably warm and dry, and the pressure to the
head suggests the idea that it may burst.   Those who
are in that very obstructed state, very seldom can use
cayenne pepper or any substance that generates much
heat, as it is not diffused and carried off by a free
perspiration, leaving the external surface cool and
comfortable.   A judicious use of means to relax the
system by the use of lobelia, white-root, crawley, sage
tea, etc., may change the conditions so that pure
stimulants may promote a natural perspiration that
will carry off the heat induced.

I will here reiterate the direction to not induce more force than what the condition of the system will appropriate and equalize.

The morbid emanations from the retained waste of an aged person may be attracted to a passive child sleeping in the same bed, while the magnetic aura, or life force of the child being relatively positive, would pass to the more aged and negative one, the law of equilibrium being constantly in force; consequently, the aged one would feel better and the child would dwindle.

In fact the laws of health, if strictly enforced, would decidedly veto the manufacture of any double beds.

LOVE AND MARRIAGE.

Deep within the human spirit is the love of the perfect and beautiful. It is so firmly established that nothing can entirely obliterate it. It is a sacred fire, burning for ever. It may sometimes be apparently smothered by antagonistic impressions which may be compared to rubbish, but it will eventually burn that rubbish and blaze with a clearer flame.

Life's manifestations in accordance with the phrenological development of an individual with large combativeness and destructiveness, without a proper restraining, refining, and overbalancing influence of the higher faculties, would be to delight in scanning the fighting qualities of a ferocious bull-dog, and would love to witness scenes of carnage and brutality. Such an individual in a republic might even attain to political eminence by the suffrages of those having similar loves.

In strong contrast is the loves of one whose intellectual and moral development is superior to the lower and more selfish faculties, because he perceives that it is lovely and just that right should be might; while the former declares that might shall be right.

Thus, what is perceived as the most lovely, attractive, and beautiful, depends on what faculties the spirit manifests through, such being the most active.

One deficient in sublimity and the higher faculties of the brain, would fail to perceive the majestic beauty and grandeur of a mountain; he would only see it as an uninteresting pile of rocks.

" Know thyself," is a direction of such superlative wisdom that it should never be forgotten.

Life manifests according to the condition of the organism; and a radiation of electric force is always active into and out of every form of matter, having polarity positive or negative, repulsive or attractive, more or less, to everything else.

Life's forces, in animate or inanimate matter, are governed by the same immutable law, unless changed by more positive mind.

The male represents the positive, the female the negative, in the general distribution of these forces.

By virtue of polarity a natural attraction exists between males and females when one is positive and the other negative. If both are positive, or both negative, they repel each other. When a male and female by virtue of their magnetic conditions are attracted sufficiently to desire a further acquaintance, the attraction grows stronger or weaker, according as they knowingly or ignorantly change the conditions.

A part of the organism which is the representative of a faculty, radiates its peculiar aura or influence which if adapted meets with reciprocation ; that makes attraction number one. If the positive magnetic aura from the representative of another faculty is also reciprocated, it is attraction number two ; and so on through the whole catalogue of the qualities of the individual.

So far the magnetic attraction or repulsion, unguided by the reasoning faculties, is common to animals and humans. Among animals, after sufficient intimacy for the positive and negative to become equalized, they are of the same polarity ; consequently they more or less repel each other.

The same effect from the same cause results from the intercourse of the human species, when these forces or causes of attraction or love are not governed and guided by the wisdom of the intellect.

Let no male or female calculate on permanency of happiness in the marriage relation if the magnetic emanations from the faculties represented by the base of the brain are the only causes of attraction.

The offspring of such union would probably be of that numerous class that are neither ornamental nor very useful, but of very selfish propensities.

Neither from a union of those whose dependence is placed too much on Providence, without the positive power to help themselves.

The offspring would be too negative and inefficient.

Neither from a union of those who are naturally positive, and without sufficient intellect to harmonize or to govern themselves.

The offspring would be of that large class who go direct from one mischief to another, and are often but erroneously termed "little devils."

Neither from a union of two positives unless the intellect predominates in both, with equal powers of reasoning to perceive truths alike, which they may equally enjoy.

And here I desire to ask pardon of the youth of both sexes, and even of older people, for advancing ideas so horridly discordant to the impressions concerning love, which have been received from novels, and from the trashy literature of the day.

Blessed is the man or woman who, after investing the object of their special regard with all the attributes which their love of the perfect and beautiful suggests as an ideal, does not feel somewhat disappointed by further acquaintance.

Again I repeat, " Know thyself;" let those of both sexes perceive the causes of attraction and repulsion, or love and hate, through what faculties life's forces are most active, and if they are liable to equalize and change polarity or repel; what resources have they in reserve for happiness in the married life. If both parties know enough to direct these forces wisely, then all may be well ; but if one, or both, does not, then it may be a failure.

ALLOPATHY.

I confess my inability to perceive the allopathic practice of medicine as a science, or as being even reasonable : but that it is based on apparently plausible but really false theories, and attribute its continuance in the market solely to its commercial value, based on the ignorance of the people concerning its really pernicious tendencies.

Instead of writing my opinion of it in many words, I prefer to quote the testimony of a few of the shining lights of the profession, presuming that they are qualified by an extensive practice of it to explain its strong points, and give proximately a righteous verdict.

M. Magendie, an eminent French physiologist, says: "I hesitate not to declare, no matter how sorely I shall wound our vanity, that so great is our ignorance of the real nature of the physiological disorders called diseases, that it would, perhaps, be better to do nothing, and resign the complaint we are called on to treat to the resources of nature, than to act, as we are frequently compelled to do, without knowing the why or wherefore of our conduct, and at the obvious risk of hastening the end of the patient."

Dr. Good says: "The science of medicine is a barbarous jargon, and the effects of our medicine on the human system are in the highest degree uncertain, except, indeed, that they have already destroyed more lives than war, pestilence, and famine combined."

"In a large proportion of cases treated by allopathic physicians, the disease is cured by nature and not by them. In a lesser, but still not a small proportion, the disease is cured by nature in spite of them; in other words, their interference opposing instead of assisting the cure."

The foregoing is quoted from an article written by

11

Dr. Forbes, M.D., F.R.S., F.G.S., Fellow of the Royal College of Physicians in London, Honorary Member of the Cambridge Philosophical Society, of the Academy of Sciences at Madrid, of the American Philosophical Society, of the Imperial and Royal Society of Physicians of Vienna, of the Royal Society of Gottingen, of the Royal Medical Society of Copenhagen, of the Medico-Chirurgical Society of Amsterdam, of the Medico-Chirurgical Society of Turin, etc., Physician in Ordinary to Her Majesty's Household, Physician Extraordinary to His Royal Highness Prince Albert, Physician in Ordinary to His Royal Highness the Duke of Cambridge, and Consulting Physician to the Hospital for Consumption and Diseases of the Chest, Editor of the Dictionary of Practical Medicine, and of the British and Foreign Medical Review.

It must be conceded at once that the forenamed is indisputable authority.

I will quote a specimen of heroic practice recommended by an English allopathic physician, the celebrated William Buchan, M.D., who published a book on medicine which had an extensive sale. In recommending remedies to produce an evacuation of the bowels in inflammation of the intestines, he says: "After other means for procuring a stool had failed, it was sometimes brought about by immersing the patient's lower extremities in cold water, or making him walk on a wet pavement and dashing his legs and thighs with cold water. In desperate cases it is common to give quicksilver; this may be given to the quantity of several ounces, or even a pound, but should never exceed that. When quicksilver is given in too large quantities, it defeats its own intention, as it drags down the bottom of the stomach, which prevents its getting over the pylorus; in this case the patient should be hung up by the heels in order that the quicksilver may be discharged by the mouth."

In the ninth edition of United States Dispensatory, published in 1851, this statement is given : " Mercury is sometimes given in the metallic state, in the quantity of a pound or two, in obstructions of the bowels, to act by its weight, but the practice is of doubtful advantage."

The use of mercury as a medicine has been very much relied on by the allopaths in almost every disorder. I will introduce the testimony of Dr. Reece, a member of the Royal College of Surgeons. He says that " The discovery of mercury was a curse instead of a blessing, since it entails more distressing diseases than it cures."

We will now notice what Dr. Brown, an eminent physician who flourished in the latter part of the 17th century, says in his published work on medicine, that he "spent more than twenty years in learning and diligently scrutinizing every part of medicine; that nothing prospered to his satisfaction, and he began to deplore the healing art as altogether uncertain and incomprehensible. All this time passed away without the acquisition of anything valuable in the healing art, and especially without that which of all things is the most agreeable to the mind, the light of truth."

(A rather discouraging verdict on the improvements in medical knowledge during all previous time.)

He finally reduced all general diseases to two forms:

" 1st. *Sthenic diathesis*—diseased habit of body occasioned by an excess of stimuli, called indirect debility, to be cured by depletion.

" 2d. *Asthenic diathesis*—diseased habit of body occasioned by a deficiency of stimuli, called direct debility, exhausted state of the system, to be cured by repletion.

" And that every power that acts on the living frame is a stimulant."

I have selected Dr. Brown's conclusions as being apparently the most reasonable of the theories of former times, and it seems to have been the guide to most of the allopathic practice to the present day. Bleeding was extensively practiced as a direct means to deplete, or destroy life's forces, thereby reducing the power to struggle against what obstructed a healthy circulation; and it was continued as long as people consented to be bled without protest.

The Nosology of Sauvages, comprises ten classes, twenty orders, three hundred and fifteen genera, and two thousand five hundred species of disease.

Dr. Good presents in his Nosology, seven classes, twenty-one orders, one hundred and thirty genera, and four hundred and eighty species.

Our distinguished countryman, Dr. Rush, declares disease to be a unit.

And Dr. Samuel Thomson, the innovator and prominent pioneer in eclecticism, says that there is one general cause of disease, viz., obstruction, to be cured by one general plan of removing it; and his practice was eminently successful.

" Who," but the people, " shall decide when doctors disagree ?"

To illustrate the allopathic, or what is falsely named the scientific practice of medicine, I will describe the treatment which General George Washington received during his last illness (from the report of his physicians) : " In the night of the thirteenth of December he was taken with a sore throat (*cynache trachealis*). He sent for a bleeder, and twelve or fourteen ounces of blood were taken. In the morning a doctor was sent for, who arrived about eleven o'clock, and who states that he then proceeded to bleed him copiously twice more ; then he was bled again, according to the account, most copiously. After this, two doses of calomel were given. Then, upon the arrival of another physician, it was agreed that as there were no signs of accumulation in the bronchial vessels of the lungs, they would try another bleeding. In addition to their previous bleedings, thirty-two ounces are now taken ;" and strange to relate, " they find that what they had already done was unattended by any alleviation of the disease. Then vapors of vinegar and water were inhaled. Ten grains more of calomel are now given, and repeated doses of emetic tartar, in all five or six grains are now administered. It is stated that the powers of life now seemed to yield to the *force of the disease*. Blisters were next applied to his extremities, and a cataplasm of bran and vinegar to his throat, to which a blister had already been applied. After many ineffectual attempts to speak, he finally expressed a desire to be allowed to die in peace. At

11*

eleven o'clock in the evening he expired," after having the foregoing amount of *medical aid*, during the period of about twelve hours.

The treatment of the foregoing case was in accordance with the theory that "excess of stimuli acting on excitability," was the cause of the disease, and that depletion was necessary. So that destroying life's forces, by bleeding and poisoning, was a direct way to cure.

It has been supposed by the unlearned and ignorant that the treatment he received would have killed a healthy man; and also that if he had used harmless, stimulating herb tea, and direct means to expand his system by warmth, to promote perspiration, equalize the circulation, and overpower the cold which he had taken, he would have conquered the cause of the inflammation in his throat.

Volumes might be written filled with the absurd theories of those who have claimed to be able to cure the sick, and larger volumes might be filled describing the detestable and unreasonable means used, which had the direct tendency to kill them.

There has been very little real progress in the allopathic practice of medicine during the past century. It has been a succession of experimenting and lauding some new preparation as almost or quite a specific for one or more symptoms, to be in its turn condemned and superseded by something else; and quite an important item of business of retail druggists at the present day is to vend secret or patent preparations claiming to be specifics, the sale depending not on

merit but on puffing by advertising. This evil would
be more tolerable if it were not for the intolerable
lying connected with it, as they are mostly advertised
as vegetable and harmless, when it is well known that
the basis of many of them is mineral and poisonous.

To illustrate the present understanding of pathology
by allopaths, I will quote from the preface to Dr.
Austin Flint's "Treatise on the Principles and Practice of Medicine," page 22, published in 1873, and
studied in medical colleges at the present day. He
says: " Pathology has been defined the study of disease : but disease has not yet been defined. The definition of disease is confessedly difficult. It is easier
to define it by negation, to say what it is not, than to
give a positive definition; that is, a definition based
either on the nature or essence of the thing defined,
or on its distinctive attributes. Disease is an absence
or deficiency of health ; but this is only to transfer the
difficulty, for the question at once arises, how is health
to be defined ? And to define health is not less difficult than to define disease."

I admire his frankness and scholastic attainments :
but how in the name of reason can the " principles
and practice of medicine" built on such foundation
be named scientific ?

I add together the items which constitute the allopathic practice of medicine, and the sum total amounts
to the following conclusions :

When life's forces struggle to circulate through an
obstructed organism, an impression is produced which,
by the nerves of sensation, the spirit perceives as dis-

order. A more powerful impression by a great variety
of means is sometimes competent to change or obliterate the symptoms, but as the quality of most of the
means relied upon by the allopaths are pernicious and
destructive, and not at all in harmony with a healthy
state, the organism is left in an obstructed condition,
with life's forces destroyed in proportion to the potency
of the means used; and generally the effects of the
medicine, as thousands can testify, are worse than the
primary disorder.

HOMŒOPATHY.

After groping through the foggy theories of the
past in the vain search for ideas in accordance with
reasonable and unchangeable principles, it is quite
refreshing to find in Hahnemann one who was able
to penetrate a little way into the world of causes.

Hahnemann was born in 1755, at Misnia, in Upper
Saxony. Educated in natural philosophy, he afterwards studied medicine, but soon became disgusted
with it as being obscure. vague, and unsatisfactory;
but by experimenting with cinchona bark on his own
person, he found it caused an intermittent fever.
Being struck with the identity of the symptoms with
the symptoms of the disease which cinchona had the
reputation of curing, he at once laid the foundation of
the doctrine of homœopathy, " *similia similibus curan-
tur*," or like cures like.

He asserts that " It is only by means of the spiritual influence of a morbific agent that our spiritual
vital power can be diseased; and in like manner, only

by the spiritual (dynamic) operation of medicine that health can be restored.

" The morbid symptoms which medicines produce in healthy persons, are the sole indications of their curative virtues in disease.

" It is a therapeutic law of nature, that a weaker dynamic affection in man is permanently extinguished by one that is similar, of greater intensity, yet of a different origin.

" It is only by the use of the minutest homœopathic doses, that the reaction of the vital power shows itself simply by restoring the equilibrium of health."

His premises I understand to be true to a certain extent.

I will quote from the " Organon " (page 217) to illustrate what the minutest doses are :

" The healing art develops for its purposes the immaterial (dynamic) virtues of medicinal substances, and, to a degree previously unheard of, by means of a peculiar and hitherto untried process. By this process it is that they become penetrating, operative, and remedial; even those that, in a natural or crude state, betrayed not the least medicinal power upon the human system.

" If two drops of a mixture of equal parts of alcohol and the recent juice of any medicinal plant be diluted with ninety-eight drops of alcohol in a vial capable of containing one hundred and thirty drops, and the whole twice shaken together, the medicine becomes exalted in energy to the first development of power, or, as it may be denominated, the first potence. The

process is to be continued through twenty-nine addi-
tional vials, each with equal capacity with the first,
and each containing ninety-nine drops of spirits of
wine ; so that every successive vial, after the first,
being furnished with one drop from the vial or dilution
immediately preceding (which had just been twice
shaken), is, in its turn, to be shaken twice, remembering to number the dilution of each vial upon the
cork as the operation proceeds.   These manipulations
are to be conducted thus through all the vials, from
the first up to the thirtieth, or decillionth '' (1,000,000,
000,000,000,000,000,000,000,000,000,000,000,000,
000,000,000,000,000,000) '' development of power,
which is the one in most general use.''

(" Organon," page 219)—" The appropriation of a
medicine to any given case of disease, does not depend
solely upon the circumstance of its being perfectly
homœopathic, but also upon the minute quantity of
the dose in which it is administered.   If too strong a
dose of a remedy that is even entirely homœopathic,
be given, it will infallibly injure the patient, though
the medicinal substance be of ever so salutary a nature ; the impression it makes is felt more sensibly,
because, in virtue of its homœopathic character, the
remedy acts precisely on those parts of the organism
which have been most exposed to the attacks of the
natural disease.''

The common dose of the solution of the thirtieth
dilution, or decillionth development of power, is one
drop, and in the dry state one globule, and generally
repeated in from one to seven days.   And a further

ͽ

direction is given, that "Homœopathic remedies operate with the most certainty and energy by smelling
or inhaling the medicinal aura constantly emanating
from a saccharine globule that has been impregnated
with the higher dilution of a medicine, and in a dry
state inclosed in a small vial, one globule, of which
10, 20, to 100 weigh a grain.

" And also, that all that is curable by homœopathy
may with the most certainty and safety be cured by
this (smelling) mode of receiving the medicine."

I have thus far quoted from Hahnemann, so as to
avoid the liability of misconstruing his strong points.
I admit that his corollaries are apparently true in theory, and practically true to a certain extent.

In the correlation of forces I admit that the greater
is capable of overcoming the less ; that the nervous
system receiving an impress from force or a modification of electricity not in accord with a healthy state,
is perceived by the spirit as disorder, pain, or ache.
A strictly homœopathic remedy which would determine to the same locality, or produce the same symptoms in a healthy person—if made magnetically potent
enough to overcome the symptoms already existing—
would relieve the patient of the primary symptoms by
substituting artificial symptoms of greater intensity,
but perhaps not as permanent.

But in any disorderly condition, life's forces are
reduced, so that waste matter is not excreted sufficiently for health, and the retained obstructive matter
is constantly emanating its disorderly forces which
determines to the weakest or most negative localities,

causing other symptoms. to be obliterated by other remedies, thus combatting effects, till the patient either dies or life's forces gain in power enough to expel obstruction, symptoms, and remedies, and an equal circulation or health is established; in which case praises are usually chanted about what is erroneously supposed to have cured them.

Hahnemann certainly deserves the thanks of the human family for having unanswerably demolished the fallacious pretentions on which allopathy is based: also, for the process of magnetizing remedies to increase their potency. And if it is possible to strictly follow his directions (which I very much doubt; certainly it is not for the average practitioner), his plan is much safer than allopathy, but is very incomplete, as it is only combatting effects or symptoms, instead of causes. And his potencies are only permanently effectual, when the obstructive causes which produced the primary symptoms have mainly spent their force in producing them.

RECIPE FOR HEALTH OF MIND.

Every individual of the human family comes into this world helpless. After being cared for until by development capacity is acquired for self-sustaining effort, each naturally takes its place as capacity and circumstances direct; but each has a natural inalienable right to as much surface of earth as is uncultivated, sufficient to supply sustenance by cultivation for the support of the physical body; and no one has the natural right to deny it except in regard to the area of land which is already occupied and cultivated, by which a right to it is previously obtained by labor on it.

But the number of acres which each may rightfully possess to cultivate should be limited. There is enough for a liberal supply for all; but some individuals are made up with a large preponderance of selfishness, and want all they can possibly get, without regard to the rights or welfare of others; those should be restrained by laws for the good of the whole.

I admit that human laws and regulations are a necessity, until all are enough intelligent and unselfish to have the law of right within themselves.

Instead of writing much on a subject which would be sure to irritate the pocket of some one, I will give a general rule for all by which to govern their actions for the good of all.

The greatest liberty which is possible to each individual, without injury to any one.

If the above simple rule is practiced by all, the existence of armies, fortifications, courts of justice, lawyers, doctors, clergymen, kings, emperors, tycoons, bailiffs, thrones, and whipping-posts, with their extravagance of expense, will become superfluous and then obsolete.

12